THE WORKS OF CHARLES DARWIN

Volume 4. *The zoology of the voyage of H.M.S. Beagle*
Part I: *Fossil Mammalia*
Part II: *Mammalia*

THE WORKS OF

CHARLES DARWIN

EDITED BY

PAUL H. BARRETT & R. B. FREEMAN

VOLUME

4

THE ZOOLOGY OF THE VOYAGE OF H.M.S. BEAGLE

PART I: FOSSIL MAMMALIA

PART II: MAMMALIA

NEW YORK UNIVERSITY PRESS
WASHINGTON SQUARE, NEW YORK

Published in 1987 in the USA by New York University Press
Washington Square, New York, NY 10003

Library of Congress Cataloging-in-Publication Data

The Zoology of the voyage of H.M.S. Beagle.

(The works of Charles Darwin ; v. 4–6)
Edited and superintended by Charles Darwin.
Includes bibliographical references and indexes.
Contents: p. 1. Fossil Mammalia ; pt. 2. Mammalia—pt. 3. Birds—pt. 4. Fish ; pt. 5. Reptiles.
1, Zoology. 2. Zoology—South America. 3. Mammals, Fossil. 4. Paleontology—South America. 5. Beagle expedition (1831–1836) I. Darwin, Charles, 1809–1882. II. Series: Darwin, Charles, 1809–1882. Works. 1987; v. 4–6.
QH365.A1 1987 vol. 4–6 575 s 87–7690
[591]
ISBN 0–8147–1790–X (pt. 1–2)

Printed in Great Britain by
The University Press, Cambridge

INTRODUCTION TO VOLUME FOUR

The Zoology of the Voyage of H.M.S. Beagle, under the Command of Captain FitzRoy, R.N., during the years 1832 to 1836. Edited and Superintended by Charles Darwin. London 1839–43. Freeman 9.[1]

Part I: *Fossil Mammalia* by Richard Owen, 1840.

During the voyage of the *Beagle*, Darwin collected large amounts of material, plants, animals and geological specimens including fossils. These he numbered, labelled, preserved and packed, sending them to England in batches as opportunity offered. He also recorded in his notebooks any relevant information. When they reached England, the specimens went to his mentor and friend John Stevens Henslow, Professor of Botany at Cambridge University.

On his return Darwin started to distribute this material amongst specialists, with a view to getting as much as possible identified, named and described in published form. He hoped to get all the vertebrate material dealt with, together with at least the molluscs and the insects. In the event, he was only able to get the former published in the *Zoology*, some of the rest being published in journals. The *Zoology* appeared in nineteen numbers, making up five parts: Fossil Mammalia, Mammalia, Birds, Fish, and Reptiles which included Amphibia, between February, 1838 and October, 1843.[2] The five parts were then issued as separate volumes from 1839 to 1843.

Richard Owen, then Hunterian Professor at the Royal College of Surgeons of England and Conservator of their Museum, undertook the mammalian fossils. Owen was the most distinguished palaeontologist in England and at that time a personal friend of Darwin: it was

[1] Freeman, R. B. *The works of Charles Darwin, an annotated bibliographical handlist*, second edition, Folkestone, 1977, p. 29.

[2] See Freeman, no. 8, op. cit., pp. 28–9 for the publication dates of the numbers.

only after the publication of *On the Origin of Species* in 1859 that they became bitter antagonists.

Darwin himself, as editor, made considerable contributions to each of the five parts. In this one he wrote a Preface of four pages as well as a ten page Geological Introduction, and Owen drew heavily on his notes throughout.

Part II: *Mammalia* by George R. Waterhouse, 1839.

George Robert Waterhouse was Keeper of Mineralogy and Geology at the British Museum, a mammalogist and entomologist. Sixty-seven species of living mammals are described here.

Darwin considered that the habits and geographical distributions were essential parts of any description of species, over and above the formal anatomical details. He insisted on these points the more strongly as his evolutionary ideas developed in the thirties, and Waterhouse used the field notes extensively.

Reproduction of the plates

The University of London have kindly given us permission to reproduce the colour plates in *Mammalia* from their copy of the original numbers. Several of the captions were altered when the numbers were rearranged for publication in book form and we have accordingly reset the captions to conform with the text set from the British Library copy of the bound volume.

The original volumes of the *Zoology* were, unlike all Darwin's other books, printed in quarto. In the present edition the illustrations have been slightly reduced, in this volume to 80 per cent of their original dimensions.

Fossil Mammalia

DESCRIBED BY

RICHARD OWEN

WITH

A GEOLOGICAL INTRODUCTION

BY

CHARLES DARWIN

THE

ZOOLOGY

OF

THE VOYAGE OF H.M.S. BEAGLE,

UNDER THE COMMAND OF CAPTAIN FITZROY, R.N.,

DURING THE YEARS

1832 TO 1836.

PUBLISHED WITH THE APPROVAL OF
THE LORDS COMMISSIONERS OF HER MAJESTY'S TREASURY.

Edited and Superintended by
CHARLES DARWIN, ESQ. M.A. F.R.S. Sec. G.S.
NATURALIST TO THE EXPEDITION.

PART I.

FOSSIL MAMMALIA:

BY

RICHARD OWEN, ESQ. F.R.S.
PROFESSOR OF ANATOMY AND PHYSIOLOGY TO THE ROYAL COLLEGE OF SURGEONS IN LONDON;
CORRESPONDING MEMBER OF THE INSTITUTE OF FRANCE, ETC. ETC.

LONDON:
PUBLISHED BY SMITH, ELDER AND CO. 65, CORNHILL.
MDCCCXL.

PREFACE

His Majesty's ship, *Beagle*, under the command of Captain FitzRoy, was commissioned in July, 1831, for the purpose of surveying the southern parts of America, and afterwards of circumnavigating the world. In consequence of Captain FitzRoy having expressed a desire that some scientific person should be on board, and having offered to give up part of his own accommodations, I volunteered my services; and through the kindness of the hydrographer, Captain Beaufort, my appointment received the sanction of the Admiralty. I must here, as on all other occasions, take the opportunity of publicly acknowledging with gratitude, the obligation under which I lie to Captain FitzRoy, and to all the Officers on board the *Beagle*, for their constant assistance in my scientific pursuits, and for their uniform kindness to me throughout the voyage. On my return (October, 1836) to England, I found myself in possession / of a large collection of specimens in various branches of natural history; but from the great expense necessary to secure their publication, I was without the means of rendering them generally serviceable.

The Presidents of the Linnean, Zoological, and Geological Societies, having given me their opinion respecting the utility to be derived from publishing these materials, I addressed a letter to the Right Honourable the Chancellor of the Exchequer (T. Spring Rice, Esq.) informing him of the circumstances under which I hoped that I might venture to solicit the aid of Government. In reply, I received a communication (as below) announcing to me that the Lords of the Treasury, from their readiness to promote science, were willing, under certain conditions, to give me the most liberal assistance.

Treasury Chambers, 31 August, 1837

SIR,

It having been represented to the Lords Commissioners of Her Majesty's Treasury, from various quarters, that great advantage would be derived to the science of natural history, if arrangements could be made for enabling you to publish, in a convenient form, and at a cheap rate, the result of your labours in that branch of science, my Lords will feel themselves justified in giving their sanction to the application of a sum, not exceeding in the whole one thousand pounds, in aid of such a publication; upon the clear and distinct understanding that the Work should be published, and the plates engraved, in such a manner as to be most advantageous to the public at large, upon a plan of arrangement to be previously submitted to, and sanctioned by the Board, after consultation with those persons, who, from their attainments / in this branch of science, are the most capable of advising their Lordships thereupon; and that the payments on account of the said sum of one thousand pounds are to be made to you from time to time, on a certificate that such progress has been made in the engravings, in accordance with the plan previously approved of, as to justify the issue then applied for. My Lords have therefore directed me to communicate to you the views they entertain upon this subject; and to apprize you that they will be prepared to act in conformity with their arrangement, upon learning from you that you are ready to proceed with the Work upon the principles above laid down, and upon receiving from you a statement of the manner in which you think the Work should be published, and the plates engraved, so as most effectually to accomplish the object my Lords have in view, in sanctioning the payment from the public funds, in aid of the expenses of the Work in question.

I remain, sir,
your obedient servant
A. Y. SPEARMAN

The object of the present Work is to give descriptions and figures of undescribed and imperfectly known animals, both fossil and recent, together with some account, in the one case, of their geological position, and in the other of their habits and ranges. As I do not possess the knowledge requisite for such an undertaking, and as I am, moreover, particularly engaged in preparing an account of the geological observations, made during the voyage, several gentlemen have most kindly undertaken different portions of the Work. Besides the very great advantage insured in thus enlisting the attainments of these naturalists in the several departments of science, to which they have paid most / attention, a great delay is avoided by adopting this method of publication, which must otherwise have been incurred before the materials could have been made known.

An account of the voyage, drawn up by Capain FitzRoy, (and to which I have added a volume) being on the point of publication, I shall not in this Work enter on any minute details respecting the countries which were visited, but shall merely give a sketch of the geology in the introduction to the part containing Fossil Mammalia, and a brief geographical notice in that attached to the account of existing animals. At the conclusion of this Work, I shall endeavour to place together the leading results in the natural history of the different countries, from which the collections were procured. I may here state that Mr Owen has undertaken the description of the Fossil Mammalia; Mr Waterhouse, the Recent Mammalia; Mr Gould, the Birds; Mr Bell, the Reptiles; and the Rev. L. Jenyns, the Fish. Whatever assistance I may obtain in the invertebrate classes, will be noticed in their respective places. The specimens have been presented to the various public museums, in which it was thought they would be of most general service: mention will be made in each part where the objects described have been deposited. /

CONTENTS

LIST OF PLATES[a]

[a] See introduction p. *6*. All the plates were drawn by G. Scharf except XVII where figs 1–4 were drawn by Laurillard. The illustrations in this octavio edition are approximately 10:8 linear reductions of the originals. Comparative sizes in captions beneath multiple illustrations have been reworded accordingly.

GEOLOGICAL INTRODUCTION

BY CHARLES DARWIN

Mr Owen having undertaken the description of the fossil remains of the Mammalia, which were collected during the voyage of the *Beagle*, and which are now deposited in the Museum of the College of Surgeons in London, it remains for me briefly to state the circumstances under which they were discovered. As it would require a lengthened discussion to enter fully on the geological history of the deposits in which these remains have been preserved, and as this will be the subject of a separate work, I shall here only give sufficient details, for the reader to form some general idea of the epoch, at which these animals lived – of their relative antiquity one to the other – and of the circumstances under which their skeletons were embedded. All the remains were found between latitudes 31° and 50° on the eastern side of South America. The localities may conveniently be classed under three divisions, namely – the Provinces bordering the Plata; Bahia Blanca situated near the confines of Northern Patagonia; and Southern Patagonia.

The first division includes an enormous area, abounding with the remains of large animals. To the eastward and southward of the great streams, which unite to form the estuary of the Plata, those almost boundless plains extend, which are known by the name of the Pampas. Their physical constitution does not vary over a wide extent; the traveller may pass for many hundred miles on a level surface, without meeting with a single pebble, or discovering any change in / the nature of the soil. The formation consists of a reddish argillaceous earth, generally containing irregular concretions of a pale brown, indurated marl. This stone, where most compact, is traversed by small linear cavities, and in several respects resembles the less pure fresh-water limestones of Europe. The concretions not unfrequently become so numerous, that they unite and form a continuous stratum, or even the entire mass.

At Bajada de Santa Fé, in the Province of Entre Rios, beds of sand, limestone, and clay of different qualities, containing sharks' teeth and sea-shells, underlie the Pampas deposit. The shells, although numerous, are few in kind. Mr George B. Sowerby informs me that they appear to belong to one of the less ancient tertiary epochs; they consist of *Venus nov. spec.* near to *V. cancellata*; *Arca nov. spec.* near to *A. antiquata*; a very large oyster, probably an extinct species; an imperfect specimen of a second species of oyster near to *O. edulis*; and a *Pecten* near to *P. opercularis*. These beds pass upwards into an indurated marl, and this again into the red argillaceous earth of the Pampas, containing the remains of those extinct quadrupeds, which every where characterize that deposit. To the southward of the Plata level plains of an uniform composition, interrupted only at wide intervals by hills of crystalline rock, extend to a distance of about three hundred miles; and to the northward for at least an equal space, and probably much further. As might have been expected from the perfectly level surface, wherever a continuous section is presented on the banks of the great rivers, very slight changes of colour show, that the deposit has been accumulated in strata as horizontal as the land, or as the water-line at the base of the cliffs.

In the province of Banda Oriental (to the N. and N.E. of the Plata), and in part of that of Entre Rios, the land, though very low and level, has a foundation of granitic and other primary rocks. These older formations are partially covered, in most parts, by a reddish earthy mass containing a few small calcareous concretions; while in other parts, they are concealed by more regular strata, of indurated marl passing into limestone, of conglomerates, and ferruginous sandstone. The entire formation probably belongs to the same epoch with that of the Pampas deposit. In the earthy mass, even where it is of little thickness, and where it might readily be mistaken for detritus produced from the underlying granites, remains of large quadrupeds have several times been discovered. /

On the shores of the Plata and in the neighbouring districts, proofs of a change of level having taken place between the land and the water within a recent period, may be observed. Both near Monte Video and Colonia del Sacramiento, beds of shells are lying on the beach at the height of several feet above the present tidal action. Near Maldonado I saw estuary shells of recent species embedded in clay, and raised above the level of a neighbouring fresh-water lake.

On the banks of the Parana, a shell identical with, or most closely resembling an estuary species (*Potamomya labiata*, now living in that part of the Plata, where the water is brackish) is accumulated in great masses, which are found some miles inland, and are elevated several yards above the level of the river. Sir Woodbine Parish, also, has in his possession, shells procured from an extensive formation near Ensenada de Barragan (south of Buenos Ayres), which is quarried for lime. Mr George Sowerby has examined these fossils, and says the following are identical with living kinds; *Voluta colocynthis*, Dillwyn: *V. angulata*, Swainson: *Buccinum globulosum*, Kiener: a variety of *Oliva patula*: a *Cytheraea* closely resembling or identical with *C. flexuosa*, and a fragment of a second species, probably *C. purpurascens*; *Potamomya labiata*; and fragments of oysters. There is, however, a species of *Mactra* in very great numbers, with which Mr Sowerby is wholly unacquainted. I may observe that I found recent shells of the first five species inhabiting the coast, a short distance to the southward. Some shelly limestone from the same place, which Sir Woodbine Parish had the kindness to show me, resembles that which I saw at Bajada, and in Banda Oriental. These beds, therefore, probably form parts of the Pampas deposit, and are not merely indicative of the period of its elevation. Nevertheless, on the opposite shores of the Plata, near the mouth of the Uruguay, I found lines of sand dunes, where the *Mactra* and *Cytheraea flexuosa* were lying in such quantities on the bare surface, that the inhabitants, by merely sifting the sand, collect them for burning into lime.

After these facts we may feel certain, that at a period not very remote, a great bay occupied the area both of the Pampas and of the lower parts of Banda Oriental. Into this bay the rivers which are now united in the one great stream of the Plata, must formerly have carried down (as happens at the present day) the carcasses of the animals, inhabiting the surrounding countries; and their / skeletons would thus become entombed in the estuary mud which was then tranquilly accumulating. Nothing less than a long succession of such accidents can account for the vast number of remains now found buried. As their exposure has invariably been due to the intersection of the plain by the banks of some stream, it is not making an extravagant assertion, to say, that any line whatever drawn across the Pampas would probably cross the skeleton of some extinct animal.

At Bajada, a passage, as I have stated, may be traced upwards from the beds containing marine shells, to the estuary mud with the bones of land animals. In another locality a bed of the same mineralogical nature with the Pampas deposit, underlies clay containing large oysters and other shells, apparently the same with those at Bajada. We may, therefore, conclude that at the period when the Arca, Venus, and Oyster were living, the physical condition of the surrounding country was nearly the same, as at the time when the remains of the mammalia were embedded; and therefore that these shells and the extinct quadrupeds probably either co-existed, or that the interval between their respective existences was, in a geological point of view, extremely short. In this part of South America there is reason to believe that the movements of the land have been so regular, that the period of its elevation may be taken as an element in considering the age of any deposit. The circumstance, therefore, that the beds immediately bordering the Plata, contain very nearly the same species of molluscs, with those now existing in the neighbouring sea, harmonizes perfectly with the more ancient (though really modern) tertiary character of the fossils underlying the Pampas deposit at Bajada, situated at a greater height, and at a considerable distance in the interior. I feel little doubt that the final extinction of the several large quadrupeds of La Plata did not take place, until the time when the sea was peopled with all, or nearly all, its present inhabitants.

Bahia Blanca, situated in latitude 39°, and about 250 miles south of the Plata, constitutes the second district, in which I found the remains of quadrupeds. This large bay is nearly surrounded by very low land, on which successive lines of sand dunes mark in many parts the retreat of the water. At some distance inland a formation of highly indurated marl, passing into limestone, forms an escarpment. Beyond this, rocks of the same character extend over a wide and / desolate plain, which rises towards the flanks of the distant mountain of the Sierra de la Ventana, composed of quartz. On the low shores of this bay, only two places occur, where any section of the strata can be seen; and at both of these I found fossil remains.

At Monte Hermoso, a line of cliff of about 120 feet in height, consists in the upper part of a stratum of soft sandstone with quartz pebbles; and in the lower of a red argillaceous earth, containing concretions of pale indurated marl. This lower bed has the same mineralogical character with the Pampas deposit; and possibly may be connected with it. The embedded bones were blackened, and had undergone more chemical change than in any other locality, which I examined. With the exception of a few large scattered bones, the remains seemed to belong chiefly to very small quadrupeds.

In another part of the bay, called Punta Alta, about eighteen miles from Monte Hermoso, a very small extent of cliff, about twenty feet high, is exposed. The lower bed seen at ebb tide, extends over a considerable area; it consists of a mass of quartz shingle, irregularly stratified, and divided by curved layers of indurated clay. The pebbles are cemented together by calcareous matter, which results, perhaps, from the partial decomposition of numerous embedded shells. In this gravel the remains of several gigantic animals were extraordinarily numerous. The cliff, in the part above high-water mark, is chiefly composed of a reddish indurated argillaceous earth; which either passes into, or is replaced by, the same kind of gravel, as that on which the whole rests. The earthy substance is coarser than that at Monte Hermoso, and does not contain calcareous concretions. I found in it a very few fragments of shells, and part of the remains of one quadruped.

From the bones in one of the skeletons, and likewise from those in part of another, being embedded in their proper relative positions, the carcasses of the animals, when they perished, were probably drifted to this spot in an entire state. The gravel, from its stratification and general appearance, exactly resembles that which is every day accumulating in banks, where either tides or currents meet; and the embedded shells are of littoral species. But from the skeleton, in one instance, being in a position nearly undisturbed, and from the abundance of serpulae and encrusting corallines adhering to some of the bones, the water, at / the time of their burial, must have been deeper than at present. This conclusion might also have been inferred from the fact, that in the neighbouring cliff the same bed, with its shells, has been uplifted some yards above high-water mark. On the coast to the southward abundant proofs occur, of a recent elevation of the continent. In the gravel, nearly all the pebbles are of quartz, and have originally proceeded from the lofty range of the Ventana, distant between forty and fifty miles. Besides the pebbles of quartz, there are a few irregular masses of the same indurated marl, of which the escarpment of the neighbouring great plain is composed. Hence the gravel beds must have been deposited, when the plain existed as dry land; and on it probably those great animals once lived, of which we now find only the remains. The indurated marl forming the plain, is the same kind of rock with that occurring over a wide extent of the Pampas; and there is no reason to doubt, they are parts of one great formation. Nevertheless, the gravel bed of Bahia Blanca, although subsequent to the calcareous formation, may be of the same age with those parts of the Pampas, which stand at a low level near the Plata. For on this whole line of coast, I believe, as the land has continued rising, fresh littoral deposits have been formed; and each of these would often owe part of its materials to the degradation of the one last elevated.

With respect to the relative age of the Monte Hermoso and Punta Alta beds, it is not possible to speak decidedly. A certain degree of similarity in the nature of the strata containing quartz pebbles, and those of the reddish indurated earth; and the short distance between the two localities, would indicate that no long interval had intervened. The beds at Monte Hermoso, certainly were deposited more tranquilly, and probably in a deeper sea; so that even skeletons of animals, no larger than rats, have been perfectly preserved there. In some parts of the surrounding country, obscure traces of a succession of step-formed terraces may be observed; and each of these indicates a period of repose during the elevation of the land, at which time the strata previously existing were worn away, and fresh matter deposited. The Monte Hermoso beds were, perhaps, formed during one such interval, anterior to the accumulation of the shingle bank at Punta Alta. /

Mr G. Sowerby, who has been good enough to examine the shells which were found with the remains of the quadrupeds, has given me the following list.

1. *Voluta angulata*
2. *Voluta colocynthis*
3. *Oliva Brasiliensis*
4. *Oliva* Nearly related to *O. patula*, but specimen imperfect
5. *Oliva* Nearly related to *O. oryza*; less nearly to small species now living at Bahia Blanca
6. *Oliva Nov. spec.*
7. *Buccinum cochlidium*
8. *Buccinum globulosum*
9. *Buccinum* One or two minute species, perhaps young specimens – unknown
10. *Trochus Nov. spec.* (?) same as one now living in the bay
11. *Trochus Nov. spec.* (?) nearly related to last; differs in not being granular on the surface
12. *Assiminia* (?) Minute species, identical with one living in the bay
13. *Bulinus nucleus*
14. *Fissurella* Probably same as a kind (*nov. spec.* ?) living in the bay
15. *Crepidula muricata*
16. *Crepidula Nov. spec.*
17. *Cytheraea* Closely related to, or identical with *C. purpurascens*
18. *Modiola* Same as recent kind (*nov. spec.*) living in the bay
19. *Nucula* Near to *N. margaritacea*
20. *Corbula* Minute species, unknown
21. *Cardita* Minute species, unknown
22. *Pecten Nov. spec.* (?) very imperfect specimen
23. *Ostrea* Oysters of the same size now live in the bay

I may add that a fossil encrusting coralline is the same with one now living in the bay.

Of these shells it is almost certain that twelve species (and the coralline) are absolutely identical with existing species; and that four more are perhaps so; the doubt partly arising from the imperfect condition of the specimens. Of the seven remaining ones, four are minute, and one extremely imperfect. If I had not made a collection (far from perfect) of the shells now inhabiting Bahia Blanca, Mr Sowerby would not have known as living kinds, five out of the twelve fossils: therefore, it is probable, if more attention had been paid to collecting the small living species, some of the seven unknown ones would also have been found in that state. The twelve first shells, as well as the four doubtful ones, are not only existing species, but nearly all of them inhabit this same bay, on the shores of which they are likewise found fossil. Moreover, at the time, I particularly noticed that the proportional numbers appeared closely similar between the different kinds – in those now cast up on the beach, and in those embedded with the fossil bones. Under these circumstances, I think, we are justified (although some of the shells are at present unknown to conchologists) in considering the shingle strata at Punta Alta, as belonging to an extremely modern epoch. /

From the principle already adduced, namely, the regular and gradual elevation of this part of the continent, I should have judged from the small altitude of the beds at Punta Alta, that the formation had not been very ancient. The conclusion here arrived at, concerning the age of these fossil mammalia, is nearly the same, with that, inferred respecting those entombed in the Pampas; and it will hereafter be shown, that some of the species are common to the two districts. We may suppose, that whilst the ancient rivers of the Plata occasionally carried down the carcasses of animals existing in that country, and deposited them in the mud of the estuary; other animals inhabited the plains round the Sierra de la Ventana, and that lesser streams, acting together with the currents of a large bay, drifted their remains towards a point, where sand and shingle were accumulating into a shoal. The whole area has since been elevated: the estuary mud of the former rivers has been converted into wide and level plains; and the shoals of the ancient Bahia Blanca now form low headlands on the present coast.

The third locality, which I have to specify, is Port St Julian, in latitude 49° 15′ on the coast of Southern Patagonia. The tertiary plains of that country are modelled into a succession of broad and level terraces, which abut one above the other; and where they approach the coast, are generally cut off by a line of precipitous cliff. The whole surface is thickly covered by a bed of gravel, composed of various kinds of porphyries, and probably originating from rocks situated within the Cordillera. The lower part of the formation consists of several varieties of sandstone, and contains many fossil shells, the greater number of which are not found in a living state.

The south side of Port St Julian is formed by a spit of flat land, of nearly a hundred feet in height; and on its surface existing species of littoral shells are abundantly scattered. The gravel is there covered (a circumstance which I did not observe in scarcely any other locality) by a thin but irregular bed of a sandy or loamy soil, which likewise fills up hollows or channels worn through it. In the largest of these channels the remains of the single fossil quadruped, which was here discovered, were embedded. The skeleton probably was at first perfect; but the sea having washed away part of the cliff, has removed many of the bones – the remaining ones, however, still occupying their proper relative position to each other. I am inclined to attribute the origin of this earthy matter, to the / mud which might have accumulated in channels, and on the surface of the gravel, if this part of the plain had formerly existed as a harbour, such as Port St Julian is at the present day. The Guanaco, the only large animal now inhabiting the wild plains of Patagonia, often wanders over the extensive flats, which are left dry at the head of the harbour during ebb tide: we may imagine that the fossil animal, whilst in a like manner crossing the ancient bay, fell into one of the muddy creeks, and was there buried.

I have stated that existing species of shells are scattered over the surface of this plain; namely, *Mytilus Magellanicus*; a second and undescribed species, now living on the beach; *M. edulis*; *Patella deaurata*; and on another part of the coast, but having similar geological relations, *Fusus Magellanicus*; *Voluta ancilla*; and a *Balanus*: all these shells are among the commonest now living on this coast. Although they must have been lying exposed to the atmospheric changes for a very long period, they still partially retain their different colours. From these facts we know, with certainty, that the superficial deposit, containing the remains of the quadruped, has been *elevated* above the sea, within the recent period. From the structure of the step-like plains, which front the coast, it is certain that each step must have been modelled, subsequently to the elevation of the one standing above it; and, as the same recent shells occur on two higher plains, we may, with safety, conclude, that the earthy matter, forming the surface of this lower one, together with its embedded skeleton, was *deposited* long after the existence of the present species, still inhabitants of the sea. According, therefore, to the chronology, taken from the duration of species among the molluscs, the fossil quadruped of Port St Julian must have been coeval, or nearly so, with those from Bahia Blanca.

Having now briefly described the principal circumstances in the geology of the three districts, to which I at first alluded, I will conclude, by observing, that the fossil mammalia of La Plata, Bahia Blanca, and Port St Julian, must all have lived during a very modern period in the geological history of the world. It is not the proper place in this work to enter on any speculations, concerning the cause of the extinction of so many gigantic animals. I will only here add, that there is the strongest evidence against admitting the theory of a period of overwhelming violence, by which the inhabitants of the land could have been swept away, and / destroyed. On the contrary every thing indicates a former state of tranquillity, during which various deposits were accumulating near the then existing coasts, in the same manner, as we may suppose others are at this day in progress. The only physical change, which we know has taken place, since the existence of these ancient mammalia, has been a small and gradual rising of the continent; but it is difficult to believe, that this alone could have so greatly modified the climate, as to have been the cause of the utter extermination of so many animals. Mr Owen will mention the exact locality where the remains of each quadruped were discovered; and, at the conclusion, it will be easy to specify by name those, which, from being embedded in the same deposit, are known formerly to have co-existed on the continent of South America. /

FOSSIL MAMMALIA

BY RICHARD OWEN

It may be expected that the description of the osseous remains of extinct Mammalia, which rank among the most interesting results of Mr Darwin's researches in South America, should be preceded by some account of the fossil mammiferous animals which have been previously discovered in that Continent. The results of such a retrospect are, however, necessarily comprised in a very brief statement; for the South American relics of extinct Mammalia, hitherto described, are limited, so far as I know, to three species of Mastodon, and the gigantic Megatherium.

One of the above species of Mastodon (*Mast. Cordillerarum*) was established by Cuvier[1] on remains discovered by Humboldt, in Quito, near the volcanic mountain, called *Imbaburra*, at an elevation of 1,200 toises above the level of the sea; and likewise at the Cordilleras of Chiquitos, near Santa Cruz de la Sierra, a locality which is near the centre of South America. A second species (*Mastodon Humboldtii*, Cuv.[2]) is indicated by molar teeth, stated to have been discovered by the same philosophic traveller, in Chile, near the city of Concepcion. The third species of Mastodon appears to have once ranged in vast troops over the wide empire of Peru: numerous teeth were brought thence to Paris by Dombey,[3] and similar teeth, together with a humerus and tibia from Santa Fé de Bogota were placed by Humboldt at the disposal of Cuvier,[4] who considered them to belong to the / *Mastodon angustidens*, a species of which the fossil remains are by no means uncommon in several localities of Europe. Cuvier is also disposed to refer to the same species the teeth of the Mastodon from

[1] See *Ossemens Fossiles*, Ed. iv, tom. ii, p. 368, Pl. 27, fig. 1, 12.
[2] Ibid. p. 370, Pl. 27, fig. 5.
[3] Ibid. pp. 347, 367.
[4] Ibid. p. 337, Pl. 26, fig. 5.

Brazil and Lima, mentioned by Dr W. Hunter in his observations on the *animal incognitum* from the Ohio.[5] The Megatherium has been scientifically described and illustrated in the works of Bru, Cuvier, and D'Alton, whose accounts are founded on a nearly complete skeleton of this stupendous quadruped which has existed in the Royal Museum at Madrid for more than half a century. The few deficiencies in its osteography have recently been supplied by the descriptions and figures given by Dr Buckland[6] and Mr Clift,[7] taken from remains of the Megatherium, brought by Sir Woodbine Parish from Buenos Ayres, and which were discovered in the bed of the Rio Salado, a tributary of the Rio Plata. Sir Woodbine Parish's collection from the same locality, includes also remains of other species of extinct Edentata, which have not yet been described. M. D'Orbigny, in his travels in South America (vol. i, p. 96), states that, in the banks of the Parana, he found the fossil remains of a large quadruped, of the size of an Ox – another quadruped of the size of a Cat, apparently of the carnivorous order; and a third, a Rodent as large as a Rat.

This meagre condition of the historical part of the subject of South American fossils by no means arises from their actual scarcity. The writings of some of the old Spanish authors, for instance, Torrubia, Garcillasso, and others,[8] contain frequent allusions to the bones of giants, who in times of old dwelt in Peru. Legentil, also, in 1728, speaks as an eye-witness of these Peruvian remains; and his guides pointed out to him the traces of the thunder-bolts, by which the Anaks of the New World had been exterminated. Bones and teeth of the Mastodon are, according to Humboldt, so abundant in a locality near Santa Fé de Bogota in Columbia, that to this day it bears the name of the 'Field of Giants'.

But independently of these indications, the abundance and variety of the osseous remains of extinct Mammalia in South America are amply attested by the materials for the following descriptions, collected by one individual, whose / sphere of observation was limited to a comparatively small part of South America; and the future traveller may fairly hope for similar success, if he bring to the search the same zeal and tact which distinguish the gentleman to whom Oryctological Science is indebted for such novel and valuable accessions.

[5] *Philosophical Transactions*, vol. lviii, p. 34 (1768).
[6] *Bridgewater Treatise*, p. 139.
[7] *Geological Transactions*, vol. iii, p. 437, Pl. 44, 45, 46.
[8] Quoted by Cuvier, *Ossem. Foss.*, Ed. iv, tom. ii, p. 351.

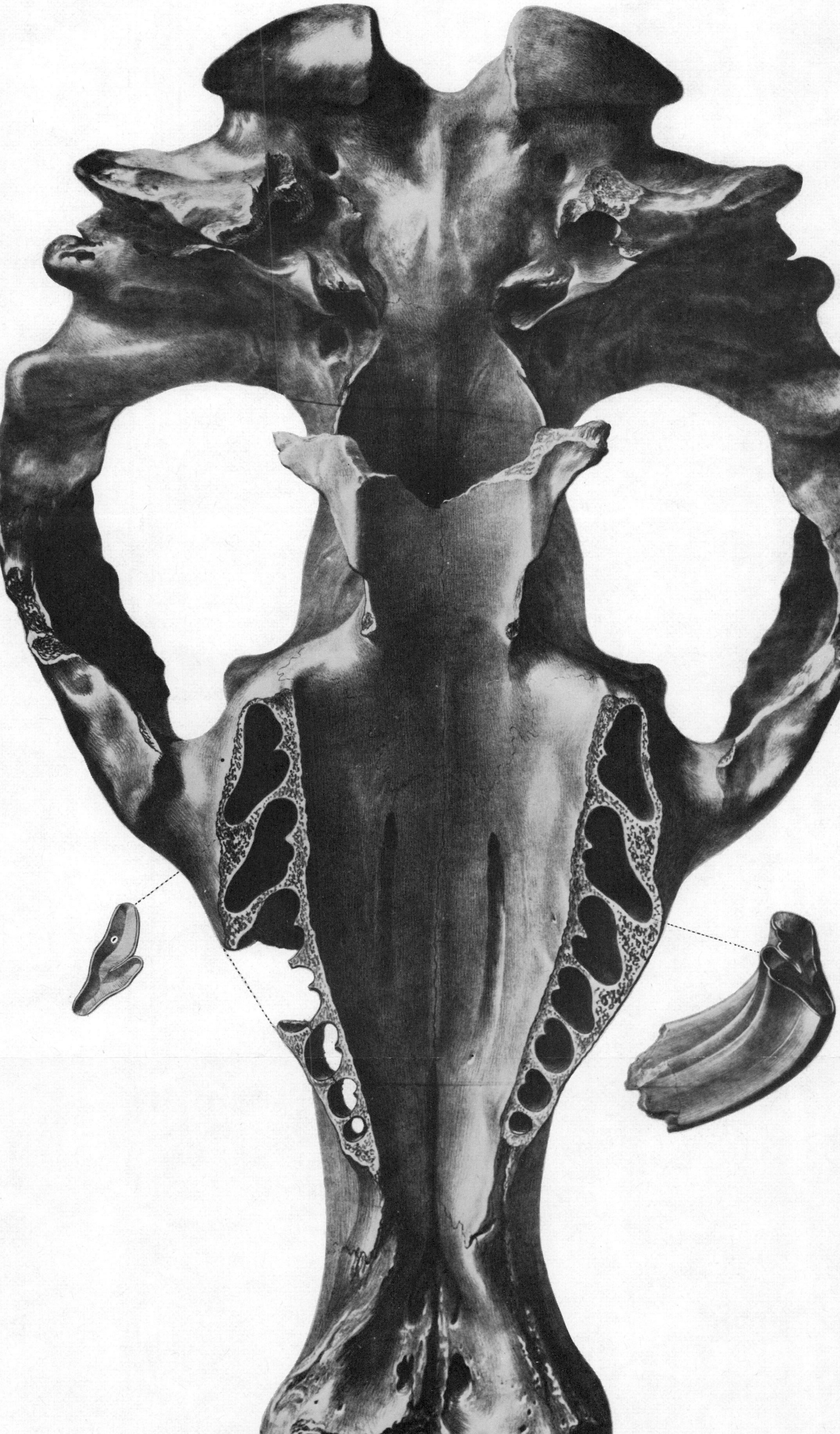

PLATE I

Base of the skull of Toxodon Platensis

It is remarkable that all the fossils, collected by Mr Darwin, belong to herbivorous species of mammalia, generally of large size. The greater part are referable to the order which Cuvier has called Edentata, and belong to that subdivision of the order (*Dasypodidae*) which is characterized by having perfect and sometimes complex molar teeth, and an external osseous and tesselated coat of mail. The Megatherium is the giant of this tribe; which, at the present day, is exclusively represented by South American species, the largest (*Dasypus Gigas*, Cuv.) not exceeding the size of a Hog. The hiatus between this living species and the Megatherium, is filled up by a series of Armadillo-like animals, indicated more or less satisfactorily by Mr Darwin's fossils, some of which species were as large as an Ox, others about the size of the American Tapir. The rest of the collection belongs, with the exception of some small Rodents, to the extensive and heterogeneous order Pachydermata; it includes the remains of a Mastodon, of a Horse, and of two large and singular aberrant forms, one of which connects the Pachydermatous with the Ruminant Order; the other, with which the descriptions in the following pages commence, manifests a close affinity to the Rodent Order. /

A DESCRIPTION OF THE CRANIUM OF

TOXODON PLATENSIS

A gigantic extinct mammiferous animal, referable to the Order Pachydermata, but with affinities to the Rodentia, Edentata, and Herbivorous Cetacea

The cranium, which is the subject of the present description, was found in the Sarandis, a small stream entering the Rio Negro, and about 120 miles to the N.W. of Monte Video: it had been originally embedded in a whitish argillaceous earth, and was discovered lying in the bed of the rivulet, after a sudden flood had washed down part of the bank.

The zoological characters deducible from this cranium, forbid its association, generically, with any known Mammiferous animal, and it must therefore be referred to an extinct genus, which I propose to call *Toxodon*,[9] from the curved or arched form of the teeth, as will

[9] Τοξόν, arcus; οδους, dens.

afterwards be described. The specific name, in the absence of other means of knowing the peculiarities of the animal than those afforded by the skull, may be most conveniently taken from the district (La Plata), in which its remains were first discovered.

The dimensions of the cranium of the *Toxodon Platensis* amply attest that the animal to which it belonged was of a magnitude attained by few terrestrial quadrupeds, and only to be compared, in this respect, with the larger Pachyderms, or the extinct Megatherium. The length of the skull (of which a base view of the natural size is given in Plate I) is two feet four inches: the extreme breadth one foot four inches. The other requisite admeasurements are given in the table at the conclusion of this description.

The general form of the skull, as seen from above, is pyriform; but viewed sideways, and without the lower jaw, it is semi-ovate; it is depressed, elongate, of considerable breadth, including the span of the zygomatic arches, but becoming rather suddenly contracted anterior to them, the facial part thence growing narrower to near the muzzle, which again slightly expands.

Among the first peculiarities which strike the observer, is the aspect of the plane of the occipital foramen, and of the occipital or posterior region of the cranium, the latter of which inclines from below upwards and forwards at an angle of 50° with the basal line of the skull. This slope of the back part of the skull is one of the characteristics of the Dinotherium; it is common to all the Cetacea, and is met with in a slighter degree in many Rodentia, and in the great Ant-eater and some others of the Edentate order. The corresponding aspect of the *foramen magnum* presents nearly the opposite extreme to man in the occipital / scale, proposed by Daubenton to determine the diversities of the form of the cranium, as a gage of the intelligence of different animals;[10] and the indication of the limited capacity of the Toxodon, thus afforded, is strengthened by the very small proportion, which the bony walls of the cerebral cavity bear to the zygomatic and maxillary parts of the skull, and to the size of the vertebral column, as indicated by the condyloid processes, and foramen magnum.

The zygomatic arches are of remarkable size and strength; they commence immediately anterior to the sides of the occipital plane, increase in vertical extent as they pass outwards, forwards and downwards, and are suddenly contracted as they bend inwards to abut

[10] *Mem. de l'Acad. des Sciences de Paris*, 1764, p. 568.

against the sides of the sockets of the two posterior molar teeth.

The cranial cavity is remarkably narrow at the space included by the zygomatic arches; being, as it were, excavated on each side to augment the space for the lodgment of the temporal muscles, so that its diameter at this part is less than that of the anterior extremity of the upper jaw. The upper surface of the cranium expands to form the post-orbital processes, and again contracts anterior to these.

The muscular ridges, or other characters, at the top of the skull, cannot be precisely determined, as a great proportion of the outer table of the bone is broken away, exposing a coarse and thick diplöe. There seems, however, to have been a strong ridge separating the occipital from the coronal or upper surface of the cranium. The form of the remaining parts, which are modified in relation to the attachment of the muscles of the jaws, indicates that these were powerfully developed both for the offices of mastication and prehension. The general form of the skull, while it presents certain points of resemblance to that of the aquatic Pachydermata, and even of the Carnivora, has much that is peculiar to itself; but, in the facial part, it approaches the nearest to that of the Rodentia; and the dentition of the Toxodon, as exhibited in the upper jaw, corresponds with that which characterizes the Rodent Order.

The teeth of the Toxodon consist of molars and incisors, separated by a long diastema, or toothless space. In the upper jaw the molars are *fourteen* in number, there being seven on each side; the incisors *four*, one very large, and one small, in each intermaxillary bone.

The general form and nature of the teeth are indicated by the sockets; and the structure of the grinders is exhibited in a broken molar, the last in the series on the left side of the jaw of the present cranium (see a figure of the grinding surface restored of this tooth, fig. 2, Pl. I), and by another perfect molar, the last but one on the right side of the upper jaw, which, though not belonging to the same individual as the skull here described, undoubtedly appertains to the same species. / This latter tooth (fig. 3, Pl. I; figs 2 and 3, Pl. IV) was found by itself, embedded in the banks of the Rio Tercero, or Carcarana, near the Parana, at the distance of a hundred and eighty miles from the locality where the head was discovered. Fragments of a molar tooth of a Toxodon, apparently the seventh of the left side, upper jaw, were also found at Bajada de Santa Fé, in the province of Entre Rios, distant forty miles from the mouth of the Rio Tercero.

All the molar teeth are long and curved, and without fangs,[11] as in most of the herbivorous species of the Rodent Order: in those, however, with curved grinders, as the *Aperea* or Guinea-pig, and *Cavia Patachonica,* the concavity of the upper grinders is directed outward, the fangs of the teeth of the opposite sides diverging as they ascend in the sockets; but, in the Toxodon, the convexity of the grinders is outward, and the fangs converge and almost meet at the middle line of the palate, forming a series of arches, capable of overcoming immense resistance from pressure. (See the upper view of the skull, Plate III, in which the fractures expose to view a part of the series of these arched sockets.)

Of the incisors, the two small ones (the sockets of which are indicated at *s s*, Pl. III) are situated in the middle of the front of the upper jaw, close to the suture between the intermaxillaries, and the two large ones in immediate contiguity with the small incisors, which they greatly exceed in size. The sockets of the two large incisors (*t t*, Pl. III) extend backwards, in an arched form, preserving a uniform diameter, as far as the commencement of the alveoli of the molar teeth: the curve which they describe is the segment of a circle; the position, form, and extent of the sockets of these incisors are the same as in those of the corresponding teeth of the Rodentia.

The matrix, or secreting pulp of the large incisors, was lodged, as in the Rodentia, in close proximity with the sockets of the anterior molars; and we are enabled to infer, from the form of the incisive sockets, notwithstanding the absence of the teeth themselves, that the pulp was persistent, and that the growth of these incisors, like those of the Rodentia, continued throughout life.

This condition, joined with the form and curvature of the socket, implies a continual wearing away of the crown of the tooth by attrition against opposing incisors of a corresponding structure in the lower jaw: and as a corollary, it may be inferred that the teeth in question had a partial coating of enamel, to produce a cutting edge, and were, in fact, true *dentes scalprarii.* The number of incisors in the upper jaw of Toxodon, is not without its parallel in the Rodent Order, the genus *Lepus* being characterized by four, instead of two superior incisors, which also present a similar relative size but have a different relative position, the / small incisors, in the hare and rabbit, being so placed

[11] True fangs exist only in teeth of temporary growth, they may be one or more in number, but always diminish in size as they recede from the crown of the tooth, and are either solid, or with a very small canal.

Toxodon, is generally almost straight, and the space included between it and the cranium is consequently narrow. The zygoma also is placed more forwards in all true Rodents, than in the Toxodon; and, instead of abutting against the posterior alveoli, it terminates opposite the anterior ones. It thus affords such an attachment to the masseter, that this muscle extends obliquely backwards to its insertion in the lower jaw, at an angle which enables it to act with more advantage in drawing forwards the lower jaw – a motion for which the joint is expressly adapted. In many Rodents, also, there is a distinct muscle, or portion of the masseter, which passes through the ant-orbital foramen, which is on that account of large size. In examining the cranium of Toxodon, with reference to this structure, it was found that the ant-orbital foramen, was not larger than might have been expected to give transmission to nerves requisite for supplying with sensibility the large lips, and whiskers with which the expanded muzzle of this remarkable quadruped was probably furnished.

Having thus examined the cranium of the Toxodon in its relation, as a / mechanical instrument, subservient to the function of digestion; we next proceed to consider the structure and composition of those cavities of the skull which gave lodgment and protection to the organs of *special* sense, and endeavour to deduce from their structure conclusions as to the degree in which the organs were developed, and the circumstances under which the senses were exercised.

The orbit of Toxodon forms the anterior boundary of the zygomatic area; it is about as distinctly defined as in the Tapir or Dugong, having its osseous rim less complete than in the Hippopotamus, yet more developed than in the Capybara, Coypus, and many other Rodentia, in which the orbit is scarcely distinguishable in the cranium from the small space occupied by the origin of the temporal muscle.

The lower boundary of the orbit in Toxodon is formed by an excavation in the upper and anterior part of the zygoma; the upper boundary by a strong and rugged overarching process of the frontal bone, the posterior angle of which (*a*, Pl. III) descends a little way, but leaves a space of three inches and a half between it and the opposite angle of the malar bone below (*b*, Pl. II and III), the circumference of the orbit being completed probably by ligament in the recent subject. The cavity thus circumscribed is remarkable for the preponderance of the vertical over the transverse or longitudinal the socket next to the posterior one of the right side. The figures subjoined of this molar tooth (fig. 3, Pl. I; figs 2 and 3, Pl. IV) almost preclude the necessity of a description. The transverse section of the tooth gives an irregular, unequal sided prism; the two broadest sides of which converge to the anterior angle, which is obtusely rounded. The / outer surface of the tooth (fig. 2, Pl. IV) is slightly concave in the transverse direction, but undulating, from the presence of two slight convex risings which traverse the tooth lengthwise. The inner surface presents at its anterior part a slightly concave surface, and posteriorly two prominent longitudinal convex ridges, separated by a groove which is flat at the bottom, and from the anterior angle of which the reflected fold of enamel penetrates the substance of the tooth, advancing obliquely forwards, rather more than half way across the body of the tooth. A longitudinal ridge of bone projects from the internal side of the socket; and fits into the groove above mentioned, and as a corresponding ridge exists in all the sockets of the grinders, save the two anterior small ones, we may infer that the five posterior grinders on each side, had a similar structure to the tooth above described. The external layer of enamel is uniformly about half a line in thickness; it is interrupted for the extent of nearly three lines at the anterior angle, and for more than double that extent at the posterior part of the tooth, which is consequently worn down much below the level of the rest of the grinding surface. Where the ivory is thus unprotected by the enamel, it has a coat of caementum, which also fills up the small interval at the origin of the reflected fold of enamel. On the grinding surface of the entire tooth, and on the fractured ends of the mutilated molars, the component fibres, or tubules, of the ivory, are readily perceptible by the naked eye, diverging from the line which indicates the last remains of the cavity of the pulp of the tooth, as it was progressively obliterated during growth.

Although the complication of the grinding surface by the inflection of simple or straight folds of enamel is peculiarly characteristic of the Rodent type, we must regard the number of molar teeth, and their diminution of size as they advance towards the anterior part of the jaw, in the Toxodon, as indicative of a deviation from that order, and an approach to the Pachyderms. The common number of grinders in the upper jaw of Rodent animals is eight, four on each side. In some genera, as Lemmus, Mus, Cricetus, there are only three on each side, and in Hydromys and Aulacodus, only two on each side. In Lepus, however, we find six on each side of the upper, and five on each side of the lower jaw. The Toxodon, like the Tapir and Hippopotamus, has seven on each side of the upper jaw: the first in each of these species being the smallest. It is worthy of notice, however, that the Capybara which adheres to the Rodent type in the number of its molars, presents in the vastly increased size, and additional number of component laminae of the posterior grinders, an approximation to the pachydermatous character just adduced, and the bony palate at the same time presents an expansion between these molars, offering a resemblance to the Toxodon which I have not found in any other Rodent besides the Capybara.

The most important deviation from the Rodent structure presented by the teeth, occurs in the direction of the reflected fold of enamel, and such a deviation / might have been inferred, even in the absence of the teeth, from the structure of the articular surface, or glenoid cavity for the reception of the condyle of the lower jaw. As the ridge of enamel runs, as above described, in a direction approaching that of the longitudinal axis of the skull, it is obvious that the grinding motions of the lower jaw should be in a proportionate degree in the transverse direction. The glenoid cavity, therefore, instead of being a longitudinal groove, and open behind, as in the true Rodents, is extended transversely, and is defended behind by a broad descending bony process preventing the retraction of the jaw, and showing marks of the forcible pressure to which it was subject.

It is worthy of observation that, in the Wombat – which exhibits the Rodent type of dentition, and, like the Toxodon, has remarkably curved molars, but in an opposite direction – the condyle of the lower jaw is also extended transversely, and adapted to an articular surface, which admits of lateral motion in the trituration of the food. In the outward span of the zygomatic arches, in which Toxodon deviates from the Rodentia, we may trace a relation of subordinacy to the above structure of the grinding teeth and joint of the lower jaw: the widening of the arches giving the masseter muscles greater power of drawing the jaw from side to side. The depth of the zygoma bespeaks the magnitude of these masticatory muscles, and the included space shews that the temporal muscles were also developed to a degree, which indicates the force with which the great incisors at the extremity of the jaws, were used; probably, like the canines of the Hippopotamus, to divide or tear up by the roots the aquatic plants, growing on the banks of the streams, which the Toxodon may have frequented.

In the Rodentia, the zygoma, though sometimes as deep as in the immediately behind the large pair, as to receive the appulse of the single pair of incisors in the lower jaw.

In the Toxodon the position of the incisors, in the same transverse line, might lead to the inference, that they were opposed by a corresponding number in the lower jaw; but the numerous examples of inequality, in the number of incisors, in the upper and lower jaws of existing mammalia, forbid any conclusion on this point.[12] The sockets of the small mesial incisors of the Toxodon (*s s*, Pl. III) gradually diminish in size, as they penetrate the intermaxillary bones, and we may, therefore, infer that the pulp was gradually absorbed in the progress of their development; and that, like ordinary incisors, their growth was of limited duration, and their lodgment in the jaw effected by a single conical fang.

I may observe, that the formation of a fang is the necessary consequence of the gradual absorption of the matrix or pulp of a tooth; for the pulp continues, as it diminishes in size, to deposit ivory upon the inner surface of the cavity of the tooth from which it is receding, and the tooth or fang thus likewise progressively diminishes in size. The formation of the socket proceeds uninterruptedly, and the bone encroaching upon the space left by the tooth, closely surrounds the wasting fang, and affords it a firm support; and thus an inference may be drawn from the form of the socket alone, as to whether the tooth it contained had or had not one or more conical fangs, and consequently whether its growth was temporary or uninterrupted.

Applying this reasoning to the molar teeth of the Toxodon, we infer that their growth, like those of most of the Phytiphagous Rodents, of the Megatherium and Armadillo, was perpetual, because their sockets are continued of uniform size from the open to the closed extremity; and the molar tooth which is preserved proves the accuracy of the deduction, inasmuch as its base is excavated by a large conical cavity for the lodgment of the pulp, the continued activity of which was the compensation here designed to meet the effects of attrition on the opposite or grinding surface of the tooth.

The molar tooth discovered by Mr Darwin in the banks of the Tercero, not only belonged to the same species as the skull under consideration, but to an individual of the same size; it fits exactly into

[12] This was written before an examination of the fragment of a lower jaw, forming part of Mr Darwin's collection of Fossil Remains, had led me to suspect that it was referable to the genus Toxodon; should this suspicion prove correct, the four unequal incisors of the upper jaw are opposed to six equal sized ones in the lower.

diameter, and indicates great extent of motion of the eyeball in the vertical direction, such as may be supposed to be well adapted to the exigencies of an amphibious quadruped. The orbit of the Capybara, or Water-hog, makes a near approach to the form just described. In the elevation of the supra-orbital boundary, and its outward projection in the Toxodon, we perceive an approximation to the form of the orbit in the Hippopotamus, but the size of the orbit is relatively larger in the Toxodon, which in this respect manifests its affinity to the Rodentia.

In that part of the bony structure of the auditory apparatus, which is visible on the exterior of the cranium, the skull of the Toxodon presents a character in which it recedes from the Rodentia. In these, the tympanic portion of the temporal bone is remarkably developed, forming a large bulla ossea between the glenoid cavity and the occiput; and it always remains disunited to the other elements of the temporal bone. In the Toxodon the tympanic bone (*c*, Pl. II) consists of a rough compressed vertical osseous plate, wedged in transversely between the occiput and the posterior part of the glenoid cavity. The internal extremity of this plate points inwards and forwards, representing the styloid process; behind this is seen the petrous bone, which forms a small angular protuberance at the basis cranii, and is less developed than in the Hippopotamus. Anterior to the petrous bone are the orifices of the Eustachian tube, and carotid canal; external to it is the great foramen lacerum, for the jugular vein and nervus vagus; and behind it is the anterior condyloid foramen. The foramen auditorium externum is only half an / inch in diameter, and gives passage to a long and somewhat tortuous meatus, which passes inwards and slightly forwards and downwards; its direction being precisely the same as in the Hippopotamus; it was accompanied, probably, by as small an external auricle.

But the indications of the aquatic habits of the Toxodon, which are presented by the osseous parts relating to the senses of sight and hearing, are of minor import compared with those afforded by the bony boundary of the nostrils. This boundary circumscribes a large ovate aperture, the aspect of whose plane is upwards, and a little forwards, as in the Herbivorous Cetaceans, and especially the Manatee (*Trichecus Manatus*, Cuv.). In one part of the bony structure of the nasal cavity the Toxodon deviates, however, in a marked degree from the Cetaceous structure; I allude to the frontal sinuses, which are exposed by the fracture of the upper part of the skull. (They are shown in Plate

III, and an asterisk is placed on one of the narrow canals of intercommunication between the sinuses and the nasal passages.) The posterior orifice of the nasal cavity is relatively larger and wider than in the Herbivorous Cetaceans, and differs both in form and aspect in consequence of the greater extent of the bony palate. The Toxodon further differs from the Manatee and Dugong, in the firm nature of the connection of the bones of the head; and it differs from the Hippopotamus in the strong attachment of the intermaxillary bones to the maxillaries.

There next remain to be described, as far as the shattered condition of the skull will permit, the relative position, extent, and connections of the principal bones composing it.

The *occipital bone* exhibits a complete confluence of its basilar, condyloid, and supra-occipital elements. The basilar portion, in connection with the corresponding element of the sphenoid bone, describes a curve whose convexity is downwards. The condyles are large, extended in the transverse direction, completely terminal, and a little inclined downwards below the level of the basilar process. The curve of the articulating surface describes, in the vertical direction, two-thirds of a circle, indicating that the head must have possessed considerable extent of motion upwards and downwards upon the atlas; thus, while the body of the Toxodon was submerged, the head probably could be raised so as to form an angle with the neck, and bring the snout to the surface of the water without the necessity of any corresponding inflection of the spine. Indeed, in the form and position of the condyles, the Toxodon more nearly resembles the true Cetacea than any other existing mammalia; and it is only with these that it can be compared in regard to the aspect of the plane of the occipital foramen, and of the occipital region of the skull. This is inclined forwards from the occipital foramen at such an angle, that on viewing the skull from above, not only the / condyles, but the entire circumference of the occipital foramen are visible. (See Pl. III.) The upper part of the supra-occipital plate presents a broad rugous depression, indicative of the insertion of strong cervical muscles, and probably of a *ligamentum Nuchae*.[13]

The ex-occipital processes advance forwards for about an inch beyond the condyles, and then suddenly extend outwards at right

[13] I have ascertained that this elastic ligament exists in the neck of the Dugong.

angles to the former line, and terminate in the form of vertically compressed bony plates; the lower rugged margins of which represent or perform the office of the mastoid processes (*d*, *d*, Pls II and III). The breadth of the entire occipital region of the skull (fig. 1, Pl. IV) appears to have been, allowing for the fractures, about one-third more than the height of the same part.

The great development of the *tympanic* bones in the Rodentia, occasions the intervention of a considerable space between the occipital bone and the zygomatic process of the temporal; but in the great Toxodon, in which the sense of hearing was doubtless inferior to that enjoyed by the small and timorous Rodents, the tympanic bone is reduced to a thin plate, which is wedged in between the occiput and glenoid cavity. In this structure, and the consequent posterior position of the glenoid cavity, there is a close resemblance between the Toxodon and the Hippopotamus, Tapir, and Rhinoceros.

The *squamous* element of the temporal bone (N, Pl. II) forms a small proportion of the lateral walls of the cranium, and also enters into the composition of the lateral and superior parts of the posterior region of the cranium, where two deep fossae perforated by large vascular foramina, indicate the junction of the squamous bones with the supra-occipital bone. The posterior surface of the skull is thus divided into three broad and shallow depressions, the two lateral facets being slightly over-lapped by the middle one, at their junction with it. In this structure the Toxodon resembles the Hippopotamus, and differs considerably from the Cetacea, in which the occipital region is rendered convex by the extraordinary development of the brain within.

The *zygomatic* process of the temporal bone projects boldly outwards at its commencement, where it is of great strength, and three-sided; the glenoid cavity extends transversely across the base or inferior surface of this part; the lateral surfaces converge to form the ridge or upper boundary of the zygoma. The depth of the glenoid cavity is increased by a transverse production of bone both before and behind it: the posterior process (*g*, Pl. II) descends the lowest, and affords the requisite defence against backward dislocation of the lower jaw; the pressure of the condyle against this process is denoted by a well defined, transversely-ovate, flattened and smooth surface, as if the bone had been planed down at that / part: the anterior transverse boundary is convex and smooth, and probably formed part of the articulation for the lower jaw. The lower facet of the zygoma anterior to the glenoid cavity gradually contracts in breadth, as it advances forward, and at the distance of three inches from the articular cavity the zygoma changes from a prismatic to a laminar form. It is at this point that the zygomatic suture commences, at the lower margin of the arch; whence it extends directly forwards for more than half its length, and then bends upwards at a right angle. The zygomatic suture has a similar course in the Capybara, and Hippopotamus.

The remainder of the zygoma is formed externally by the *malar* bone (G Pl. II), which in its position is intermediate to the Rodent and Pachydermatous structures. It is not suspended in the middle of the zygomatic arch, as in the former order; neither does it extend into the region of the face so far anterior to the orbit as in the Tapir or Hippopotamus. The exterior line of the malo-maxillary suture defines the orbit anteriorly; but from this line the maxillary bone extends backwards, along the inner side of the malar portion of the zygoma, until it almost reaches the temporo-malar suture; thus abutting by an oblique surface against nearly the whole internal facet of the malar bone, and materially contributing to the general strength of the zygomatic arch. The malar bone is of considerable vertical extent, and presents a rugged and thickened inferior margin for the attachment of the masseter. The upper margin of the malar bone is smoothly rounded, and presents a regular semi-circular excavation, forming the lower boundary of the orbit. The relative magnitude of the zygomata to the entire cranium far exceeds in the Toxodon that which exists in the Hippopotamus or any other known Pachyderm. This arises from the great vertical development of the malar bone behind the orbit, and the vertical expansion of the temporal portion of the arch. The oblique position of the zygoma, descending as it advances forwards, is deserving of attention, as the Toxodon, in deviating from the Pachyderms in these respects, makes an evident approach to the herbivorous Cetaceans, as the Dugong and Manatee: in the latter Cetacean we observe a similar development of the lower part of the zygomatic process of the malar bone. It is here, also, that we may perceive an indication of a resemblance between the Megatherium and Toxodon.

There is no discernible trace of the *lachrymal* bone (E, Pl. II), having extended, as in the Hippopotamus beyond the anterior boundary of the orbit: the lachrymal foramen is situated rather deep in the orbit, and the bone itself appears to have been of very small size.

The surface of the supra-orbital processes of the *frontal* bone (C, Pl.

PLATE II

Side view of the skull of Toxodon

II) is deserving of attention, as it presents a peculiar ruggedness which is not found in any other part of the skull; the irregularity seems, as it were, to have been produced by the impression of numerous small tortuous and anastomosing vessels. In the / skull of a Sumatran two-horned Rhinoceros, in the Museum of the College of Surgeons (No. 816), the circumference of that part of the surface of the skull which supported the posterior horn, and which includes precisely the same part of the os frontis, presents the same character, the surface being broken by numerous vascular impressions. On the supposition that this character of the supra-orbitary arch in the Toxodon might indicate the superincumbency of a bony case, I examined the skulls of two Armadillos, *Dasypus Peba* and *Das. 6-cinctus*, and found that in the Dasypus 6-cinctus, the supra-orbital ridges, which are slightly elevated, to support the cephalic plate, presented, in a minor degree, a corresponding rugosity. May we venture then to conjecture that the Toxodon was defended by an ossified and integument like the Armadillo, or that it was armed with an epidermic production, analogous to the horn of the Rhinocerous; or had the rugous surface in question as little relation with the parts that covered it as the sculptured surface of the malar bones in the Cavy?

After forming the rugged and prominent supra-orbital processes already described, the frontal bone continues to send backwards a slightly elevated ridge or *crista*, circumscribing the origin of the temporal muscles, but the extent of this ridge, and the disposition of the inter-orbital portion of the frontal bones cannot be determined in the present mutilated specimen. The fractures it has sustained are not, however, wholly unattended with advantage; they expose the structure of the diploë, which from its coarseness of texture and thickness, resembles that of the Cetaceous crania; and what is of still more importance, they also demonstrate the existence and form of the frontal sinuses.

The cavity of the nose is extensive, and the remains of the ossa spongiosa superiora testify that the Toxodon enjoyed the sense of smell to a degree equal at least to that of the Hippopotamus.

The *sphenoid bone* resembles that of the Hippopotamus, but it contributes a larger share to the formation of the internal pterygoid processes (*p*, Pl. II); these are of a simple form, and more developed than in the Hippopotamus; they project outwards to a greater extent, and terminate in a point. The sphenoid also sends off a short and thick pointed process from the posterior part of the base of the internal pterygoid processes. The ala of the sphenoid does not rise so far into the orbit, nor does it articulate with the parietal bone, as in the *Hippopotamus*; but in this part of its structure, is the same as in the Rhinoceros. The spheno-palatine foramen is relatively larger than in the above-named Pachyderms, and is bounded above by the descending orbital plate of the frontal bone.

The palatal processes of the *palatine* bones terminate anteriorly between the last molars, and extend backwards for some distance beyond the alveolar processes, increasing the extent of the bony roof of the mouth posteriorly: this is a structure in which the Toxodon deviates both from the Rodents, and Pachyderms, / and resembles the Armadillos, among the Edentata; excepting that the post-dental part of the bony palate in the Toxodon is suddenly contracted in breadth. The palato-maxillary suture is in the form of a chevron, with the angle directed forwards, as in the Hippopotamus and Capybara, but truncated.

The *superior maxillary* bones (F, Pl. II) are united posteriorly to the malar, as above described: they ascend and join the frontal and nasal bones: their outer surface is almost vertical, smooth, and slightly undulating; perforated at its posterior part by the ant-orbital foramen, and joined anteriorly to the intermaxillaries by a suture running in the sigmoid direction (as shewn in Pl. II) from the middle of the nasal cavity, to within four inches of the anterior boundary of the upper jaw. We have, in the position and extent of this suture, and the absence of tusks and their large prominent sockets, a most important difference between the Toxodon and Hippopotamus. The chief peculiarity in the maxillary bones, obtains in the arched form of the alveolar processes, corresponding to the shape and position of the grinders above described, and which are peculiar among known mammalia to the present genus. The palatal surface of the maxillary bones is obliquely perforated by two large foramina, from which two deep longitudinal grooves extend forwards, and are gradually lost; we find the posterior palatine foramina represented by similar grooves and foramina in the Capybara.

The *intermaxillary* bones (D, Pls II and III), though large, are relatively of less extent than in the Rodents generally. The nasal processes do not reach the frontal bone, but are limited to the anterior half of the nasal boundary; approaching in this respect to the Herbivorous Cetacea. In the outward expansion of their anterior extremities, the intermaxillaries resemble those of the Hippopotamus, in which, however, this character is more strongly marked. The intermaxillaries in the Hippopotamus are also much less firmly united to the maxillary bones than in the Toxodon, and are consequently commonly lost in the fossil crania. On the palatal surface of the intermaxillary bones there are two grooves which diverge forwards from the line of the suture; and anteriorly to these grooves there are the two large anterior palatine foramina. The maxillo-intermaxillary sutures on the palate converge as they extend backwards to a point; there appears to have been a fissure left between this suture and the mesial suture of the intermaxillaries; in which structure the Toxodon resembles the Hippopotamus.

After summing up the different affinities, or indications of affinity, which are deducible from the cranium of this most curious and interesting fossil mammal, we are led to the conclusion, assuming it to have had extremities cased in hoofs, that it is referable to the Order Pachydermata. But the structure, form, and kind of teeth in the upper jaw, prove, indisputably, that the gigantic Toxodon was intimately related to the Rodent Order. From the characters of this order, as afforded by the existing species, the Toxodon, however, differs in the relative / position of the supernumerary incisors, and in the number, and direction of the curvature, of the molars. If, moreover, the lower jaw, next to be described, belong, as I believe, to the Toxodon, the dental character of the genus will be *incisors* $\frac{4}{6}$; *pro laniariis diastema*; *molares* $\frac{7}{7}\,\frac{7}{7}$.

The Toxodon again deviates from the true Rodentia, and resembles the Wombat, and the Pachyderms, in the transverse direction of the articular cavity of the lower jaw.

It deviates from the Rodentia, and resembles the Pachydermata in the relative position of the glenoid cavities and zygomatic arches, and in many minor details already alluded to.

In the aspect of the plane of the occipital foramen, and occipital region of the skull; in the form and position of the occipital condyles; in the aspect of the plane of the anterior bony aperture of the nostrils; and in the thickness and texture of the osseous parietes of the skull, the Toxodon deviates both from the Rodentia and existing Pachydermata, and manifests an affinity to the Dinotherium and Cetaceous Order, especially the Herbivorous section.

At present we possess no evidence to determine whether the extremities of the Toxodon were organized on the ungulate or unguiculate type, nor can we be positive, from the characters which

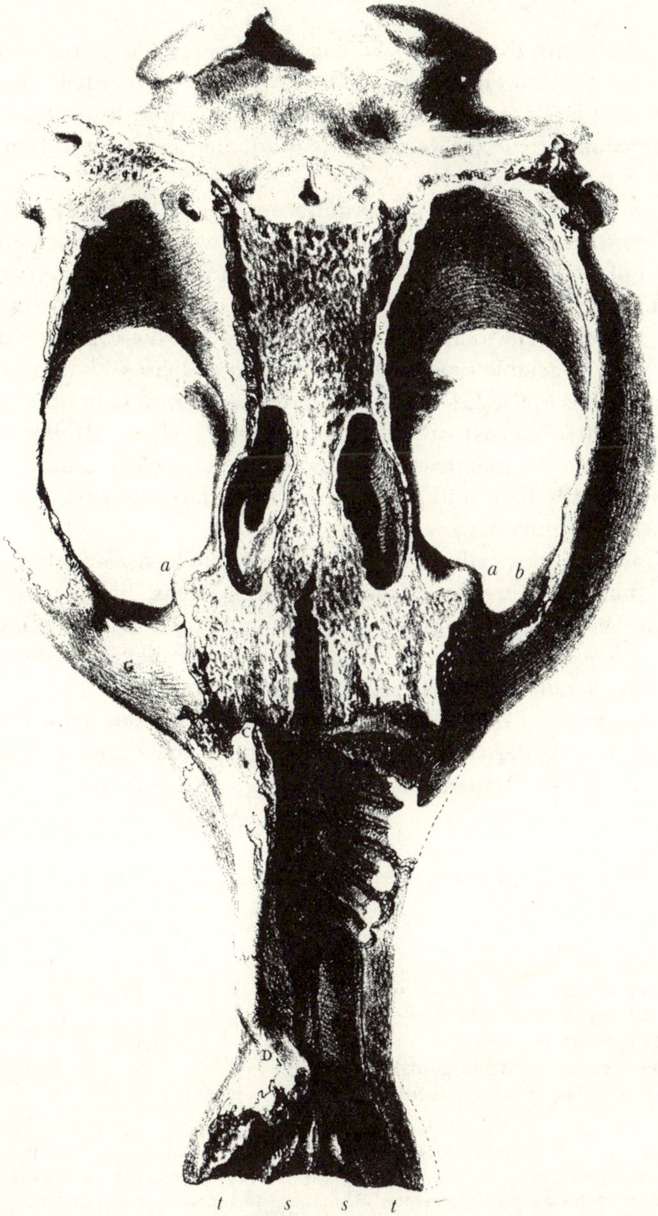

PLATE III

Top view of the skull of the Toxodon

the skull affords, that the genus may not be referable to the *Mutica* of Linnaeus;[14] although the development of the nasal cavity and the presence of large frontal sinuses render it extremely improbable that the habits of this species were so strictly aquatic, as the total absence of hinder extremities would occasion.

Where the dentition of a mammiferous animal is strictly carnivorous, this structure is obviously incompatible with a foot incased in a hoof: but where the teeth are adapted for triturating vegetable substances the case is different. If animals so characterized are of small size and seek their food in trees, or if they burrow for roots or for shelter, the vegetable type of dentition must co-exist with unguiculate extremities, as in the Edentata and Rodentia generally: but the largest genus (Hydrochaerus) of the Rodent Order, whose affinity to the Pachydermata is manifested in its heavy shapeless trunk, thinly scattered bristly hair, and many other particulars, has each of its toes enclosed in a miniature hoof.

The affinity above alluded to, is too obvious to have escaped popular notice, and the Capybara, from its aquatic habits, has obtained the name of Water-hog. It is highly interesting to find that the continent to which this existing aberrant / form of Rodent is peculiar, should be found to contain the remains of an extinct genus, characterized by a dentition which closely resembles the Rodent type, but manifesting it on a gigantic scale, and tending to complete the chain of affinities which links the Pachydermatous with the Rodent and Cetaceous Orders.

ADMEASUREMENTS OF THE CRANIUM OF TOXODON

	Feet	*Inches*	*Lines*
Extreme length	2	4	–
Extreme breadth	1	4	–
Extreme height (exclusive of the lower jaw)	–	10	–
Length of zygomatic process	1	1	6
Depth or vertical extent of zygomatic process	–	6	–
Transverse extent of zygomatic fossa	–	6	–

[14] The German Translator (See *Frorieps Notizen.*, 1837, p. 119) of the abstract of my description of the Toxodon, published in the *Proceedings of the Geological Society*, asks, what is the *Mutica* (misprinted *Muticata*), of Linnaeus? The term is quoted from the *Systema Naturae*, Ed. xii, p. 24. Linnaeus first divides Mammalia into three groups, according to modifications of the locomotive organs, viz. *Unguiculata*, *Ungulata*, *Mutica*, and subdivides these, according to modifications of the dentary organs, into the orders, *Bruat*, *Glires*, *Primates*, etc.

ADMEASUREMENTS OF THE CRANIUM OF TOXODON – *continued*

	Feet	*Inches*	*Lines*
Transverse diameter of cranium between the zygomatic arches	–	5	–
Transverse diameter of occipital plane of the cranium	1	–	–
From the outside of one condyle to that of the opposite condyle	–	8	6
Length of the bony palate	1	6	–
Extreme breadth of the bony palate	–	6	–
Breadth of palate at the intermaxillary suture	–	2	6
Breadth of palate behind the molar alveoli	–	3	–
Longitudinal extent of the molar alveoli	–	9	6
Longitudinal extent of the diastema	–	5	6
Transverse diameter of posterior nasal aperture	–	3	9
Transverse diameter of occipital foramen	–	3	–
Transverse diameter of glenoid cavity	–	4	6
Antero-posterior diameter of glenoid cavity	–	1	–

DESCRIPTION OF FRAGMENTS OF A LOWER JAW AND TEETH OF A TOXODON

Found at Bahia Blanca, in latitude 39° on the East coast of South America

In looking over some fragments of jaws and teeth, forming part of Mr Darwin's collection of South American mammiferous remains, and which had been set aside with mutilated specimens referable to species belonging to the family of Edentata, my attention was caught by the appearance of roots of teeth projecting, in a different direction from the grinders, from the fractured anterior extremity of a lower jaw, and I was induced to examine minutely the structure of the teeth in this specimen, and to search the collection for corresponding fragments. The result was the discovery of portions of the two rami, and the commencement of the symphysis of a lower jaw, containing anteriorly the roots of / six incisors, and at least six molars on each side; but as the rami had been fractured through the middle of the sixth alveolus, the number of grinders may have corresponded with those in the upper jaw of the Toxodon.

The most perfect of these fragments is figured in Pl. V, figures 1 and 4; figure 2 shows the form of the teeth in transverse section, and the disposition of the enamel upon the grinding surface of the molars on

Fig. 1

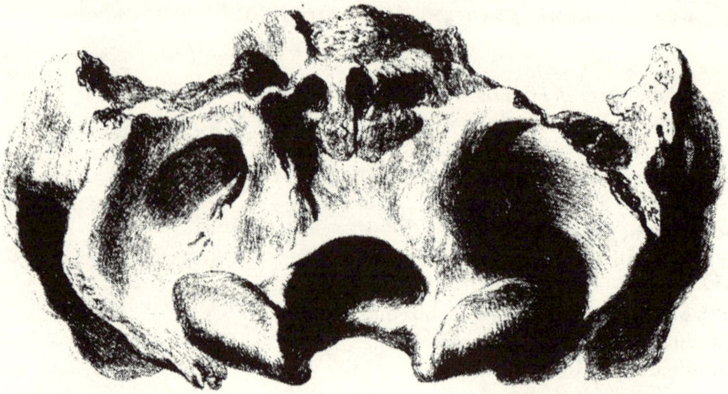

Fig. 2

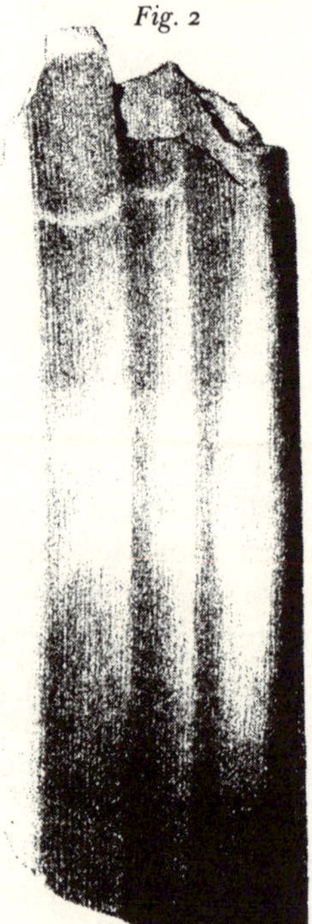

Fig. 3

PLATE IV

Fig. 1. Toxodon Platensis *Figs 2, 3. Sixth grinder upper jaw*

Fig. 1 is drawn ⅓ of the size of Figs 2 and 3

the right side, as restored from a comparison of the fractured teeth in the two rami. From the remains of the symphysis shown at fig. 4, it will be seen that the jaw was remarkably compressed, or narrow from side to side; while the rami (fig. 1) were of considerable depth, in order to give lodgment to the matrices and bases of grinders enjoying uninterrupted growth.

The pulps of the six incisors in this lower jaw are arranged in a pretty regular semi-circle, whose convexity is downwards; the teeth themselves are directed forwards, and curved upwards, like the inferior incisors of the Rodentia. The form and degree of the curvature are shown in the almost perfect incisor (Pl. V, fig. 5) which corresponds with the left inferior incisor of the lower jaw, and was found in the same stratum, but belonged to another individual.

These incisors are nearly equal in size: they are all hollow at their base, and the indurated mineral substance impacted in their basal cavities well exhibits the form of the vascular pulps which formerly occupied them. Sufficient of the tooth itself remains in four of the sockets to show that these incisors, like the nearly perfect one (fig. 5), had only a partial investment of enamel; but though in this respect, as well as in their curvature and perpetual growth, they resemble the dentes scalprarii of the Rodentia, they differ in having a prismatic figure, like the inferior incisors of the Sumatran Rhinoceros, or the tusks of the Boar. Two of the sides, viz., those forming the anterior convex and mesial surfaces of the incisor have a coating of enamel, about half a line in thickness, which terminates at the angles between these and the posterior or concave surface. In plate V, fig. 4, the enamel of the broken incisors is represented by short lines, showing the direction of its crystalline fibres; the white space immediately within the enamel shows the thickness of the ivory at the base of the tooth, the included grey substance represents a section of the formative matrix or pulp of the tooth, which was of the usual conical form: the inferior broken end of the incisor (fig. 5) appears to have been distant about one-third from the apex of the pulp.

From the relative position of the bases or roots of these incisors, we may infer that they diverged from each other as they advanced forwards, in order to bring their broadest cutting surfaces into line. That they were opposed to teeth of a corresponding structure in the upper jaw is proved by the oblique chisel-like cutting surface of the more perfect incisor: and it is not without / interest to find that the presence of *dentes scalprarii* at the anterior part of the mouth has not been necessarily limited to Mammalia of small size.

The position of the pulps of these incisors, in close proximity with the anterior grinders, corresponds with the position of the pulps of the incisors in the upper jaw of the *Toxodon*, and indicates, in conjunction with the size of the pulps, that a considerable extent of the inferior incisors was lodged in the substance of the anterior part of the jaw. It is most likely that no vertically directed tooth would be developed in the part of the jaw so occupied by the curved bases of the incisors, and hence a diastema or toothless space would intervene between the molars and incisors of this lower jaw, as in the upper jaw of the Toxodon.

It is interesting, also, to observe, that as the deviations from the Rodent type, which occur in the cranium of the Toxodon, are the same, in some instances, as those which obtain in the Wombat; so we find a corresponding deviation in the size and relative position of the inferior incisors, which, as in the Wombat, terminate anterior to the molar teeth, instead of extending backwards beyond the last grinder, as in most of the true Rodents. The Capybara presents the nearest approach to this structure, the pulps of the inferior incisors being situated opposite the interspace of the first and second grinders.

The molar teeth, in this mutilated lower jaw, like those in the upper jaw of Toxodon, had persistent pulps, as is proved by the conical cavity at their base, as represented in fig. 3; they consequently required a deep socket, and a corresponding depth of jaw to form the socket and protect the pulps. In order to economize space, and to increase the power of resistance in the tooth, and perhaps, also, to diminish the effects of direct pressure on the highly vascular and sensible matrix, we find the molars and their sockets are curved, but in a less degree than those of the upper jaw of the Toxodon. They correspond, however, with the superior molars of the Toxodon in the antero-posterior diameter, in being small and simple at the anterior part of the jaw, and by increasing in magnitude and complexity as they are situated more posteriorly. They are, however, narrower from side to side; but supposing them to belong to the Toxodon, it would agree in this respect with most other large herbivorous mammalia; the fixed surface for attrition in the upper jaw being from obvious principles more extensive than the opposed moveable surface in the lower jaw.

The *first* grinder, in the lower jaw here described (Pl. V, fig. 2), is of small size and simple structure, being surrounded with a coating of enamel of uniform thickness, and without any fold penetrating the substance of the tooth. It is more curved than any of the other molars, and appears to have differed from the external incisor only in its entire coating of enamel and direction of growth; it is interesting, indeed, to find so gradual a transition, in structure, from molar to incisive teeth, / as this jaw presents; for the robust incisors may here be regarded as representing molars simplified by the partial loss of enamel, and with a change in their direction.

In the *second* molar, we find an increase in the antero-posterior diameter, and in the length of the tooth, and the enamel at the middle of the outer side makes a fold which penetrates a little way into the tooth; the line of enamel, on the inner side, is slightly concave and unbroken.

The *third* molar presents an increase of dimensions in the same directions as the second; the enamel on the outer side of the tooth presents a similar fold, but it is directed a little more backwards.

In the *fourth* molar, besides a further increase of size, and a corresponding but deeper fold of enamel in the external side of the tooth, we have the grinding surface rendered more complicated by two folds of enamel entering the substance of the tooth from the inner side: these folds divide the antero-posterior extent of the tooth into three nearly equal parts; they are both directed obliquely forwards, half-way across the substance of the ivory.

The *fifth* molar presents the same structure as the fourth, which it exceeds only slightly in size.

In the *sixth* molar we have a proportionally greater increase of size in the antero-posterior diameter, which measures two inches; but the lateral diameter is but slightly augmented; its structure resembles that of the fifth.

As these grinding teeth by no means increase in the lateral diameter in the same proportion as in their antero-posterior diameter, the posterior ones present, but in a greater degree, the compressed form which characterizes the grinders of the upper jaw of the Toxodon.

It will be seen, however, that there is a difference in the structure of the grinders in this fragment of the lower jaw and those of the upper jaw of the Toxodon. In the lower grinders there are two folds of enamel proceeding from the inner side of the tooth into its substance, whilst in the upper grinders there is only one fold continued from the inner side; in the lower grinders there is also a fold of enamel reflected into the substance of the tooth from the outer surface, while in the upper grinders of Toxodon we find the enamel coating on the outer side of the tooth merely bent inwards, so as to describe, in the

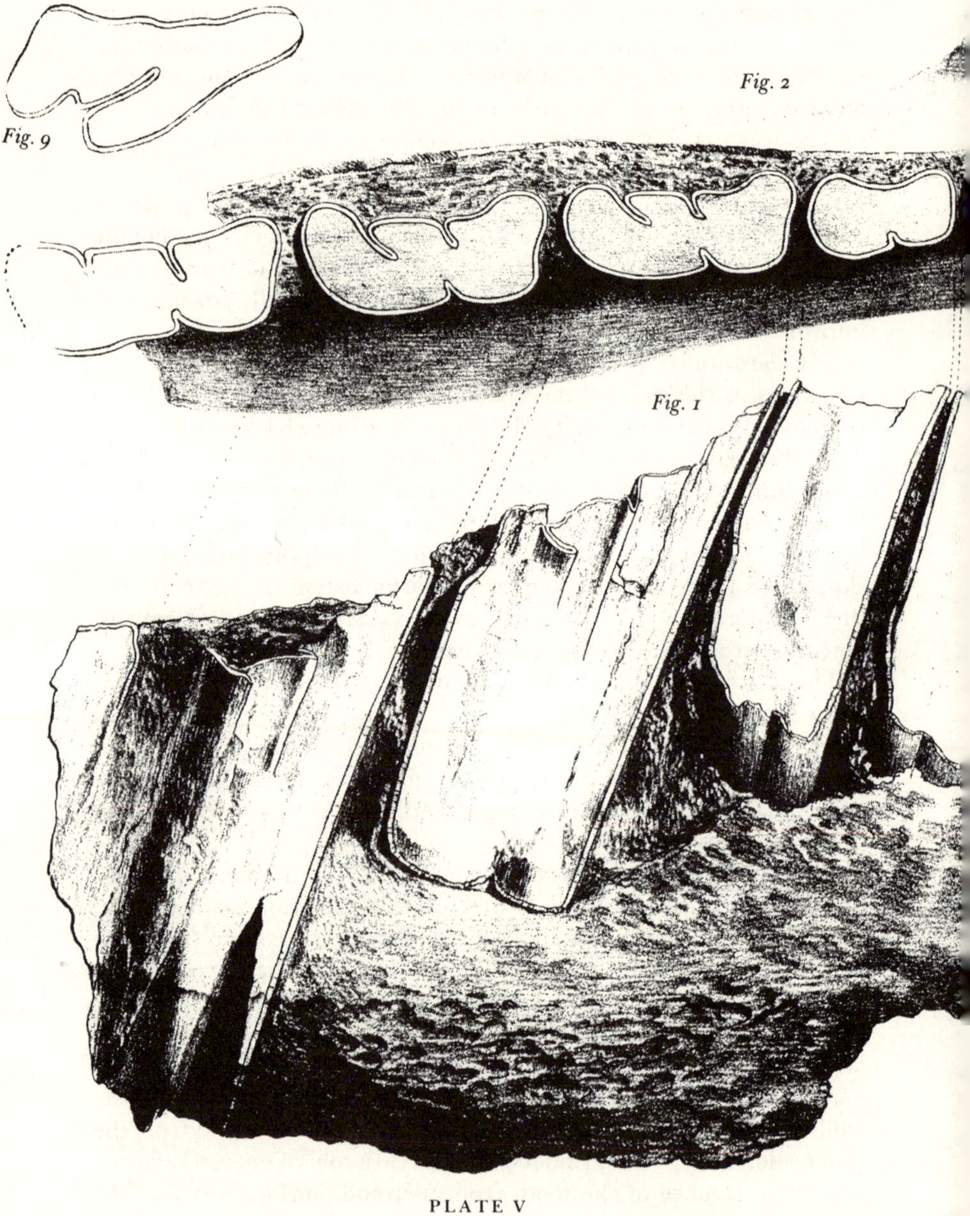

PLATE V

Fragments of the lower jaw and teeth of a Toxodon

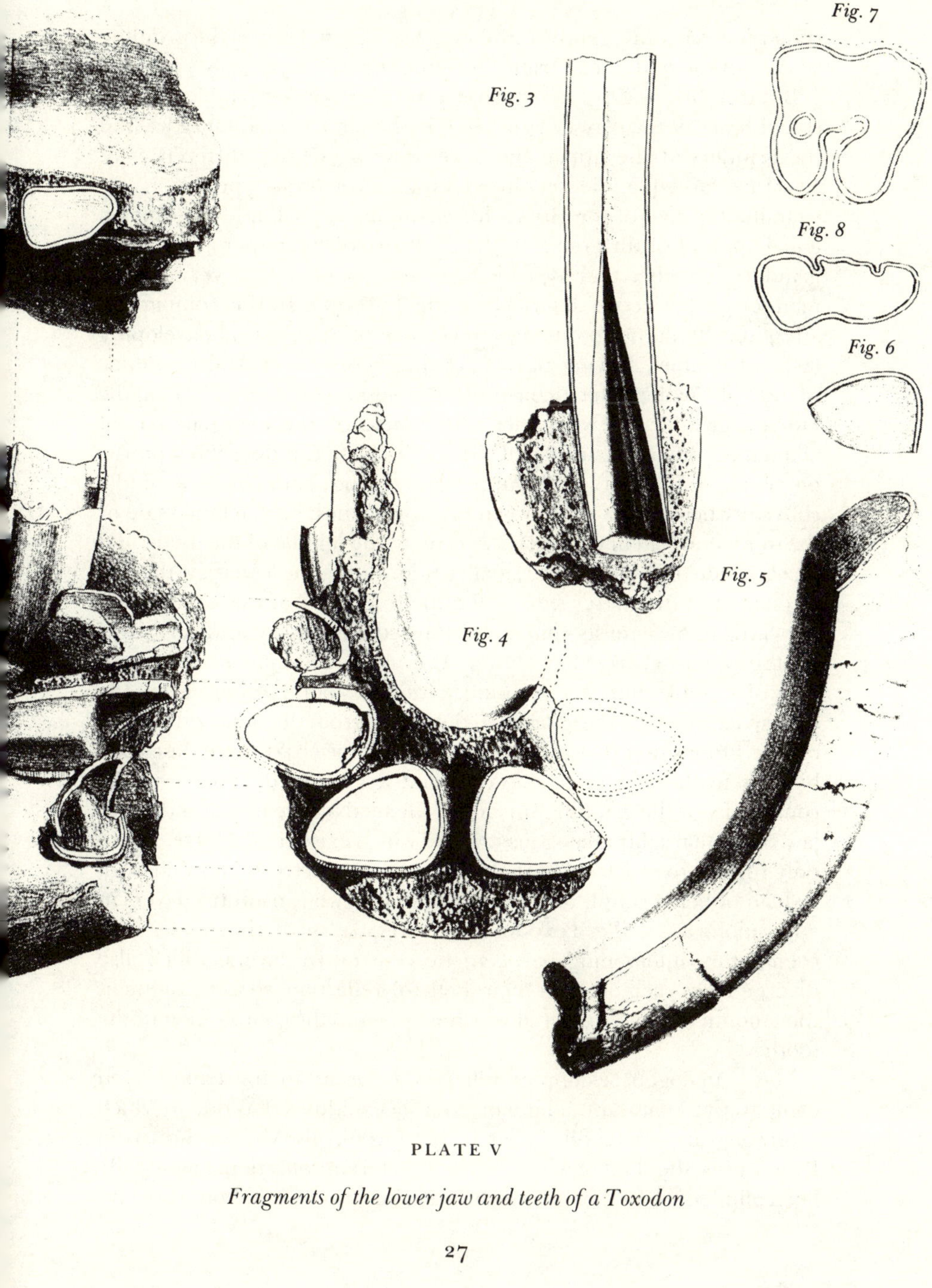

PLATE V

Fragments of the lower jaw and teeth of a Toxodon

transverse section, a gently undulating line; fig. 7, Pl. V, is the grinding surface of the sixth molar, right side, upper jaw.

But this difference of structure is by no means incompatible with the co-existence of the two series of teeth in the same animal, since we find the grinders of the upper and lower jaws presenting differences of structure of equal degree in existing herbivorous species. If we examine the jaws of the Horse, for example, we shall find not only an equal amount of difference in the structure of the upper / and lower grinders, but that they deviate from one another in a very similar manner to that above described in the Toxodon. In this comparison attention should be confined to the course of the external enveloping layer of enamel, leaving out of consideration the central crescentic islands of enamel which constitute the additional complexity of the Horse's grinder. Viewing then the course of the external coat of enamel on the worn surface of the tooth, we find it describing on the outer side of the tooth in the upper jaw an undulating line – a middle convexity being situated between two concavities; on the inner side of the tooth one fold of enamel penetrates to the middle of the tooth, and on each side of this there is a smaller fold. But in the lower jaw the line of enamel on the outer side of the tooth, instead of merely bending outwards midway in its course, is reflected a little way inwards; while on the opposite, or inner side of the tooth, the enamel sends two extensive folds into the substance of the tooth, opposite to the interspace of which the shorter fold projects from the outer side. Now, on the supposition that the fragment of the lower jaw here described belongs to the Toxodon, the kind and degree of difference in the complexity of the grinding surface of the teeth in the upper and lower jaw, are remarkably analogous to those which exist in the Horse. I have only further to remark that in the Horse the inflected folds of enamel, instead of being simple and straight with the two constitutive layers in apposition, as in the Toxodon, are irregular in their course, with coementum intervening between the constitutive layers, which also diverge from each other at their angle of reflection, so as to augment the amount of dense material which enters into the composition of the tooth.

Many analogous examples will readily occur to the experienced comparative anatomist. The Horse has been adduced as one to which reference can very readily be made; but I would also cite the Sumatran Rhinoceros, the skull of which, in the Hunterian collection, has already been alluded to. In this species the anterior grinders, in both jaws, are small and simple, and increase in complexity as they recede backwards. The third superior grinder (fig. 8, Pl. V) presents a single fold of enamel, reflected obliquely forwards from the inner side half-way across the tooth; the outer line of enamel describes a simply undulating line. The opposite grinder of the lower jaw (fig. 9, Pl. V) has only one-half the breadth of the upper one, but has its grinding surface further complicated by having two inflected folds of enamel from the inner side, and one shorter and broader fold from the outer side. This tooth, therefore, presents a close resemblance to one of the posterior grinders of the lower jaw of the Toxodon, but differs essentially in being of limited growth, and consequently in having fangs.[15] /

In speculating upon the nature of the organized substances which the teeth of the Toxodon were destined to grind down, we must not only take the structure of the tooth into consideration, but also the power of perpetual renovation, which will compensate for the defective quantity of enamel in the grinders of the Toxodon, as compared with those of the existing Ruminants and Pachyderms, whose grinders, when once completed, receive no further addition of dental substance at their base. The Toxodon, in this character of its dentition, participated in the same advantages with the Capybara and the Megatherium.

Although we have been enabled to observe the structure of the grinding teeth of the upper jaw of the Toxodon in two examples only; one, an insulated perfect grinder corresponding to the sixth alveolus on the right side, and the other, a portion of the last grinder of the left side remaining in the socket of the head previously described, yet from the relations subsisting between socket and tooth, a very satisfactory opinion may be formed of the structure of those teeth which are wanting, as well as of their size. It thus appears, that the grinders of the upper jaw of the Toxodon, are small and simple at the anterior part of the jaw, and that they increase (chiefly in antero-posterior extent) in size, as well as in complexity, as they recede backwards in the jaw. In this respect, as well as in size, the teeth, in the fragments of the lower

[15] Besides the relation to *food requiring much comminution,* which teeth with persistent pulps bear, they are also connected with the *longevity of the individual.* The term of life in a herbivorous animal, with grinders / of temporary growth, is, of necessity, dependent on the duration of these essential aids to nutrition; thus, a sheep generally wears down its grinders in twelve years, and its natural term of life is consequently limited to about that period.

jaw just described, exactly correspond. There is, however, a slight difference in the lateral diameter of the two sets of grinders, those of the lower jaw being narrower, as is usually the case, but not in the same degree as in the Horse or Ruminant. A greater difference obtains in the degree of curvature of the two sets of molars, those of the lower jaw, especially the posterior grinders, being much less bent than the corresponding teeth of the upper jaw. It is necessary to observe, also, that the convexity of the curve of the inferior grinders is directed outwards, as in the superior grinders; while in the Guinea Pig and Wombat, which have also curved grinders, the convexity is outwards in the lower jaw, and inwards in the upper jaw.

Nevertheless, if we take into consideration the close similarity which exists between the teeth of the upper jaw of the Toxodon, and those of this lower jaw in more essential points, as in their persistent pulps, their characteristic structure and form, the depth of their sockets, and their relative sizes and complexity; and when we consider how the depth of this lower jaw, and its narrowness in the transverse direction, corresponds with the characteristic form of the upper jaw of the Toxodon, and that to these resemblances is added an apparatus of incisors adequate to oppose the great dentes scalprarii of the upper jaw, the conclusion seems irresistible, / that the lower jaw, here described, must be referred, if not to the same, at least to a nearly allied species of Toxodon, as that to which the large cranium belonged.

Further researches in South America, it is hoped, will lead, ere long, to the completion of our knowledge of the osteology of this very remarkable and interesting genus of extinct mammiferous animals.

DESCRIPTION OF PARTS OF THE SKELETON OF

MACRAUCHENIA PATACHONICA

A large extinct Mammiferous Animal, referable to the Order Pachydermata; but with affinities to the Ruminantia, and especially to the Camelidae

In the preceding pages the nature and affinities of a large extinct Mammal were attempted to be determined from the cranium and teeth exclusively: we come now to consider the remains of a quadruped consisting of bones of the trunk and extremities, without a fragment of a tooth or of the cranium to serve as a guide to its position in the zoological scale.

It may appear, even to anatomists and naturalists familiar with the kind of evidence afforded by a fossil fragment, that an opinion as to the relation of the present species to a particular family of Ruminants, formed without a knowledge of the important organs of manducation, must be vague and doubtful, but the evidence about to be adduced, will be regarded, it is hoped, as more conclusive than could have been *à priori* expected.

The portions of the skeleton of the animal – which, in relation to the affinity above alluded to, as well as from the length of its neck, I propose to call *Macrauchenia*[16] – were discovered by Mr Darwin in an irregular bed of sandy soil, overlying a horizontal accumulation of gravel on the south side of Port St Julian: and independently of the circumstances under which they were found, their correspondence with each other in size, colour, texture and general character prove them to have belonged to one and the same individual.

These remains include two cervical vertebrae, seven lumbar vertebrae, all more or less fractured; a portion of the sacrum and ossa innominata; fragments of the left scapula; of the right radius and ulna, and right fore-foot; the right femur / nearly entire, the proximal and distal extremities of the right tibia and fibula; and a metatarsal bone of the right hind foot.

Before entering upon the description of these remains, a few observations may be advantageously premised on some of the distinguishing characters of the Camelidae. It is well known that the Camels and Llamas deviate in their dentition, viz., in the presence of two incisors in the upper jaw, from the true Ruminants; and we cannot avoid perceiving that in this particular the direction in which they deviate tends towards the conterminous Ungulate Order, in which incisor teeth are rarely absent in the upper jaw. They also further deviate from the Ruminants and approach the Pachyderms in the absence of cotyledons in the uterus and fetal membranes; having, instead thereof, a diffused vascular villosity of the chorion, as in the sow and mare.

But besides these characters, by which, in receding from one type of

[16] Μακρος longus, αυχην *cervix*: from the latter word Illiger derived *Auchena*, his generic name of the Llama, Vicugna, etc.

hoofed mammalia, the Camelidae claim affinity with another, there are many parts of their organization peculiar to themselves; of some of these peculiarities, the relation to the circumstances under which the animal exists, can be satisfactorily traced; in others, the connection of the structure with the exigencies of the species, is by no means obvious, and in this predicament stands the osteologial peculiarity, which is immediately connected with our present subject – a peculiarity in which the Camelidae differ not only from the other Ruminants, but from all other existing Mammalia, and which consists in the absence of perforations for the vertebral arteries in the transverse processes of the cervical vertebrae, the altas excepted.

I may observe that what is described as a perforation of a single transverse process in a cervical vertebra is essentially a space intervening between two transverse processes, a rudimental rib, and the body of the vertebra. In the cold-blooded Saurians – in which the confluence of the separate elements of a vertebra takes place tardily and imperfectly, if at all – the nature of the so called perforation of the transverse process is very clearly manifested, as in the cervical vertebrae of the Crocodile, in which the interspace of the inferior and superior transverse processes is closed externally by a separate short moveable cervical rib. In the Ornithorhynchus paradoxus the vertebra dentata also preserves throughout life this condition of its lateral appendages: in other Mammalia it is only in the foetal state that the two transverse processes are manifested on each side with their extremities united by a distinct cartilage, which afterwards becomes ossified and anchylosed to them.

In the Hippopotamus the inferior transverse process sends downwards a broad flat plate extended nearly in the axis of the neck, but so obliquely, that the posterior margins of these processes, in one vertebra, overlap the anterior ones of the succeeding vertebra below, like the cervical ribs in the Crocodile; the same structure obtains in many other mammalia, especially in the Marsupials. In the / Giraffe, the inferior transverse processes are represented by relatively smaller compressed laminae, projecting obliquely downwards and outwards from the anterior and inferior extremity of the body of the vertebra. The superior transverse processes in this animal are very slightly developed in any of the cervical vertebrae, and the perforation for the vertebral artery is above and generally in front of the rudiment of this process, being continued as it were through the side of the substance of the body of the vertebrae.

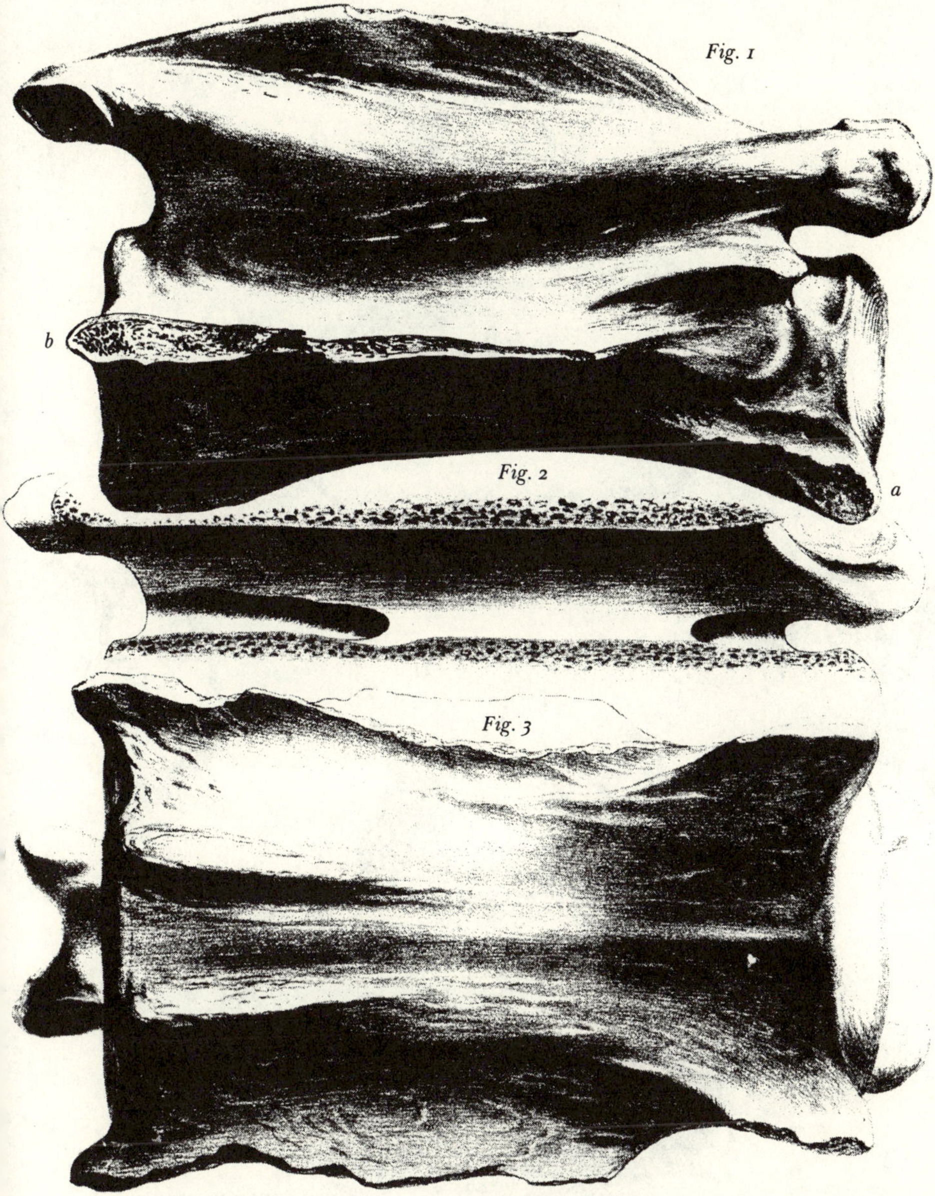

PLATE VI

Cervical vertebrae of Machrauchenia

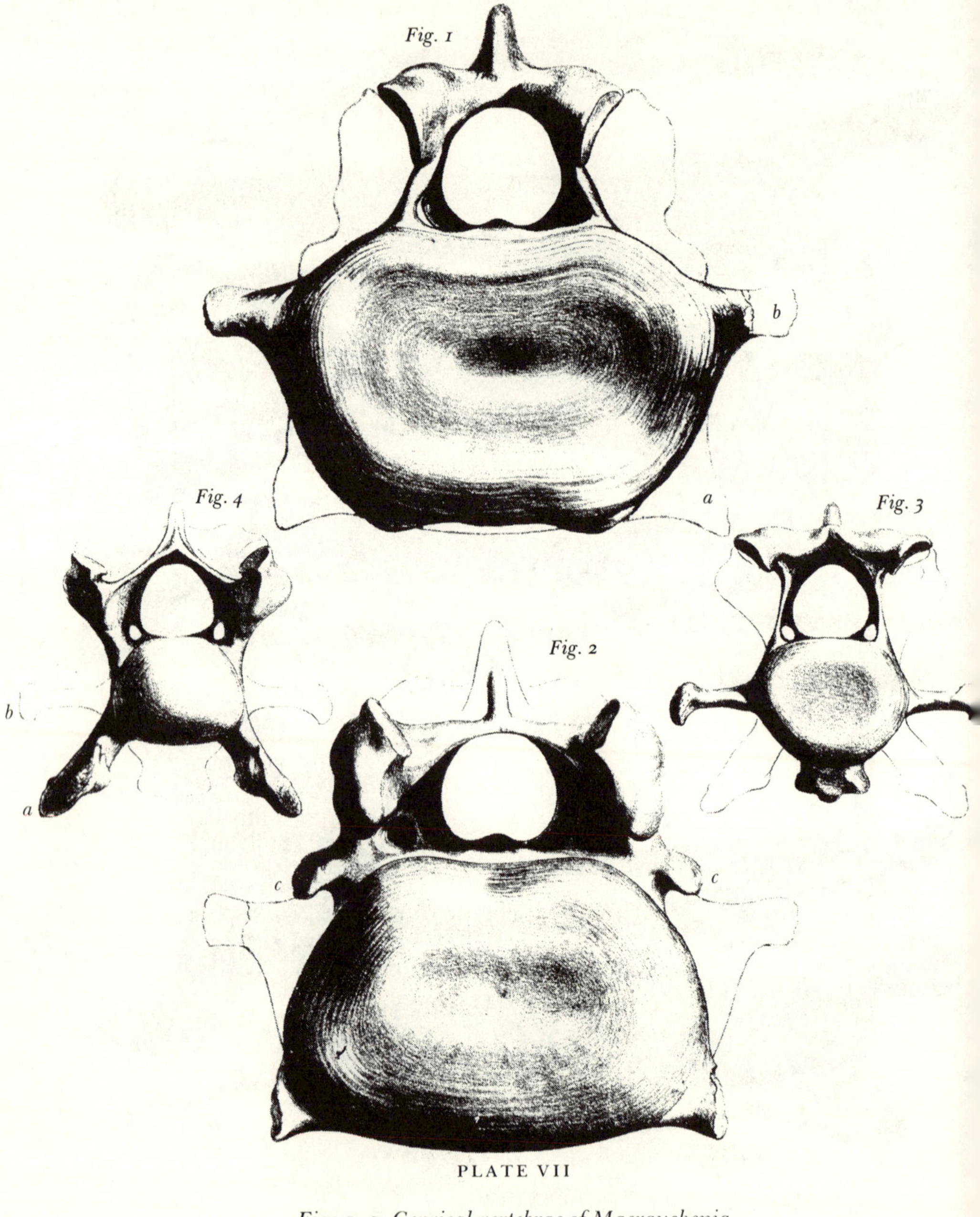

PLATE VII

Figs 1, 2. Cervical vertebrae of Macrauchenia
Figs 3, 4. Cervical vertebrae of Auchenia

In the long cervical vertebrae of the Camel and Llama, the upper and lower transverse processes are not developed in the same perpendicular plane on the sides of the vertebrae, but at some distance from each other; the lower transverse processes (*a*, fig 1, Pl. VI; *a*, fig. 1, 3, 4, Pl. VII) being given off from the lower part of the anterior extremity of the body of the vertebra; the upper ones (*b*, fig. 1, Pl. VI; *a*, fig. 1, 3, 4, Pl. VII) from the base of the superior arch near the posterior part of the vertebra, or from the sides of the posterior part of the body of the vertebrae. The extremities of these transverse processes do not become united together, but they either pass into each other at their base, or continue throughout life separated by an oblique groove (as in fig. 1, Pl. VI). This groove would not, however, afford sufficient defence for the important arteries supplying those parts of the brain which are most essential to life; and, accordingly the vertebral arteries here deviate from their usual course, in order that adequate protection may be afforded to them in their course along the neck. From the sixth to the second cervical vertebrae inclusive in the *Aucheniae*, and from the fifth to the second inclusive in the *Cameli*,[17] the vertebral arteries enter the vertebral canal itself, along with the spinal cord, at the posterior aperture in each vertebra, run forwards on the outside of the dura mater of the cord between it and the vertebral arch, and when they have thus traversed about two-thirds of the spinal canal, they perforate respectively the superior vertebral laminae, and emerge directly beneath the anterior oblique or articulating processes, whence they are continued along with the spinal cord into the vertebral canal of the succeeding vertebra, and perforate the sides of the anterior part of the superior arch in like manner; and so on through all the cervical vertebrae until they reach the atlas, in which their disposition, and consequently the structure of the arterial canals, resemble those in other Ruminants.

The two cervical vertebrae of the Macrauchenia present precisely the structure / and disposition of the bony canals for the vertebral arteries which are peculiarly characteristic of the Camelidae among

[17] In the seventh cervical vertebra of the Camel, as in many other Mammalia, there is no perforation in any part for the vertebral arteries. In a Vicugna, I find the same structure; but in a Llama, the side of the body of the seventh cervical vertebra is perforated longitudinally on the right side. In the Camel, the vertebral arteries pierce the sixth cervical vertebra, immediately below the superior transverse processes, and pass obliquely to the anterior aperture of the cervical canal, where they emerge beneath the anterior oblique processes, and then enter the spinal canal of the fifth cervical vertebra, as described in the text.

existing Mammalia. In Plate VI, fig. 2, the groove and orifices of the canal for the vertebral artery are shown in a section exposing the spinal canal: in Plate VII, figures 1 and 3 exhibit the orifices at the commencement of the arterial canals, as seen in a posterior view of the vertebrae; in figs 2 and 4, the terminations of the same canals are shown, in the anterior view of the same vertebrae; the smaller figures (3 and 4) are taken from the fourth cervical vertebra of a Llama. The vertebrae of the Macrauchenia also closely resemble the middle cervical vertebrae of the Vicugna and Llama in their elongated form; approaching the Auchenial division of the Camelidae, and deviating from the true Camels in the relations of the length of the body of the vertebra to its breadth and depth, and in the much smaller size of the inferior processes. Excepting the Giraffe, there is no existing mammal which possesses cervical vertebrae so long as the Macrauchenia; but the cervical vertebrae of the Giraffe, differ in the situation of the perforations for the vertebral arteries, and in the form of the terminal articular surfaces, as will be presently noticed.

Both of the cervical vertebrae of the *Macrauchenia* here described, are of the same size, each measures six inches and a half in extreme length, two inches, ten lines in breadth, and two inches, four lines in depth. In the Giraffe and the Camelidae, the spinous processes are thin laminae of considerable extent in the axis of the vertebra, but rising to a very short distance above the level of the vertebral arch: the spinous processes have the same form in the corresponding vertebrae of the Macrauchenia, but present a still greater longitudinal extent; they commence at the interspace of the anterior oblique processes, and extend to opposite the base of the posterior oblique processes; the upper margin describing a gentle curve, as shown in fig. 1, Pl. VI. The transverse processes also present the form of slightly produced, but longitudinally extended, laminae: their disposition is essentially the same as in the Camelidae, but more nearly corresponds with the modifications presented by the Aucheniae. The inferior transverse processes – those which are alone developed in fish, but which are not present in any other vertebrae save the cervical, in mammalia – these processes in the Macrauchenia are continued from the sides of the under surface of the anterior part of the body of the vertebra; their extremities being broken off, it cannot be determined how far they extended from the body of the vertebrae, but they gradually subside as they pass backwards: the superior transverse processes are continued outwards from the sides of the posterior part of the body of the vertebra, and gradually subside as they advance forwards along three-fourths of the body of the vertebra: they are not continued into the anterior and inferior transverse processes, as in the Vicugna, but are separated therefrom by a narrow and shallow groove. The articular, or oblique processes, closely resemble those of the Auchenia / in form, and in the direction of the articular surfaces; those of the anterior processes looking inwards and a little upwards; those of the posterior, outwards and a little downwards.

In the Macrauchenia a small longitudinal process (*c*, fig. 2, Pl. VII) is given off immediately below the base of the anterior oblique process; this structure is not observable in any of the cervical vertebrae of the Giraffe or Camelidae.

In the form of the articulating surfaces of the bodies of the vertebrae the Macrauchenia deviates from the Giraffe and Camel, but resembles the Aucheniae. In the Giraffe and Camel the anterior articulating surface is convex and almost hemispheric, the posterior surface is proportionally concave, so that the cervical vertebrae are articulated by ball and socket joints; yet not, as in most Reptiles, with intervening synovial cavities, but by the concentric ligamentous intervertebral substance characteristic of the Mammiferous class. In the Llama and Vicugna, the degree of convexity and concavity in the articular surface of the bodies of the cervical vertebrae is much less than in the Camels; and in consequence they carry their necks more stiffly and more in a straight line. In Macrauchenia the anterior articulating surface (fig. 2, Pl. VII) presents a still slighter convexity than in the Llama (fig. 4, Pl. VII), and the posterior surface (fig. 1, Pl. VII) presents a correspondingly shallower concavity. The form of the extremities of the body of the vertebrae, especially of the posterior, is sub-hexagonal, the breadth being to the depth as eight to five. The sides and under part of the vertebrae are slightly concave; on the inferior surface there are two ridges, continued forwards from the posterior margin of the vertebra, each situated about an inch distant from the middle line; they converge as they pass forwards, and are gradually lost in the level of the vertebra; their greatest elevation does not exceed half an inch. In the Aucheniae there is a longitudinal protuberance in the mesial line, instead of the two ridges. The two long cervical vertebrae of the Macrauchenia are also characterized by the maintenance of an almost uniform diameter of the body, both in its vertical and transverse extent; the cervical vertebrae of the Vicugna come nearest to them in this respect; those of the Camel deviate further in the large excavation at the under part of the body.

The long vertebral or spinal canal offers a slight enlargement at the two extremities; this structure which is generally in the ratio of the extent of motion of the vertebrae on each other is more marked in the Camel, where the form and mode of articulation of the bodies of the vertebrae are designed to admit of a free and extensive inflection of the cervical vertebrae; and the result of this structure is very obvious in the sigmoid flexure of the neck in the living animal. In the Aucheniae, on the contrary, the neck is carried less gracefully erect and in an almost straight line, and the form of the vertebrae and the nature of their joints correspond, as we have seen, to this condition. From the length of the bodies of the / cervical vertebrae of the Macrauchenia, and the almost flattened form of their anterior and posterior articular surfaces, I infer that the long neck in this singular quadruped must have been carried in the same stiff and upright position as in the Vicugna and Guanaco.

The following individual differences are observable in the two cervical vertebrae of the Macrauchenia; in the posterior one the superior arch is wider and with thicker parietes, the body is more concave below, and the inferior transverse processes have a more lengthened origin.

Not a fragment of dorsal vertebrae, ribs or sternum, is included in the collection of the bones of the Macrauchenia; but fortunately seven lumbar vertebrae, forming a consecutive series of the same individual as that to which the cervical vertebrae belonged, were obtained, all more or less fractured, but all sufficiently perfect to demonstrate their true nature. These vertebrae, although not possessing such distinctive characters as the cervical, contribute by no means an unimportant element towards the illustration of the osteology of the Macrauchenia, and support the view which I have taken of its affinities; for, although, as will be seen from the structure of its extremities, this animal must be referred to the Order Pachydermata, yet no existing species of that order has more than six lumbar vertebrae; whilst among the Ruminants it is only in the Camel, Dromedary, Llama and Vicugna, that the lumbar vertebrae reach the number seven – the same number which characterizes the extinct annectant species in question. The dimensions of the vertebrae in the Macrauchenia present the same relations to the two cervical vertebrae above described, which the lumbar vertebrae of the Vicugna bear to the third, fourth, or fifth of its cervical vertebrae. But here we begin to discover modifications of form, in which the Macrauchenia deviates from the Camelidae, and approaches the Pachyderms, as the Horse and Hippopotamus; and these indications become stronger as the vertebrae approach the sacrum.

In the Camel, as well as in the Horse and Hippopotamus, the bodies of the lumbar vertebrae diminish in vertical extent, or become flatter, as they approach the sacrum; but this character is more strongly marked in the Macrauchenia than in either of the above species. But in the Camelidae the transverse processes of the lumbar vertebrae, are elongated, flattened, and narrow, resembling ribs, except that they are nearly straight; and this is more particularly the case with the transverse processes of the last lumbar vertebrae, which are the narrowest of all in proportion to their length, and stand freely out without touching the sacrum. The transverse processes of the lumbar vertebrae of the Giraffe resemble those of the Camel, but are relatively smaller and shorter. In the Hippopotamus the transverse processes of the lumbar vertebrae are much broader in proportion to their length than in any of the Ruminants, and they increase in breadth to the / last lumbar vertebra, which presents in addition, the following characters; each transverse process sends off from its posterior margin a thickened and transversely elongated protuberance, which supports a flattened articular surface adapted to a corresponding surface on the anterior part of the transverse process of the first sacral vertebra: it likewise presents on its anterior edge a flattened and rough surface, which is closely attached by ligamentous substance to the opposite part of the transverse process of the penultimate lumbar vertebra. A similar structure exists in the last two lumbar vertebrae of the Rhinoceros, Tapir, and Horse. In the latter animal anchylosis of these articulating surfaces of the lumbar and sacral vertebrae generally takes place with age, and, judging from the character of the same surfaces in the Hippopotamus, the motion of its lumbar vertebrae upon the sacrum may in like manner become ultimately arrested.

Now in the Macrauchenia, as in the Pachyderms above cited, the transverse processes of the last lumber vertebrae are of considerable thickness and extent, and are joined by enarthrosis to the transverse processes of the sacrum; but the bony structure of these joints would indicate that they were not subject to be obliterated by anchylosis. The articular surfaces which project from the posterior part of the transverse processes of the last lumbar vertebrae present a regular and

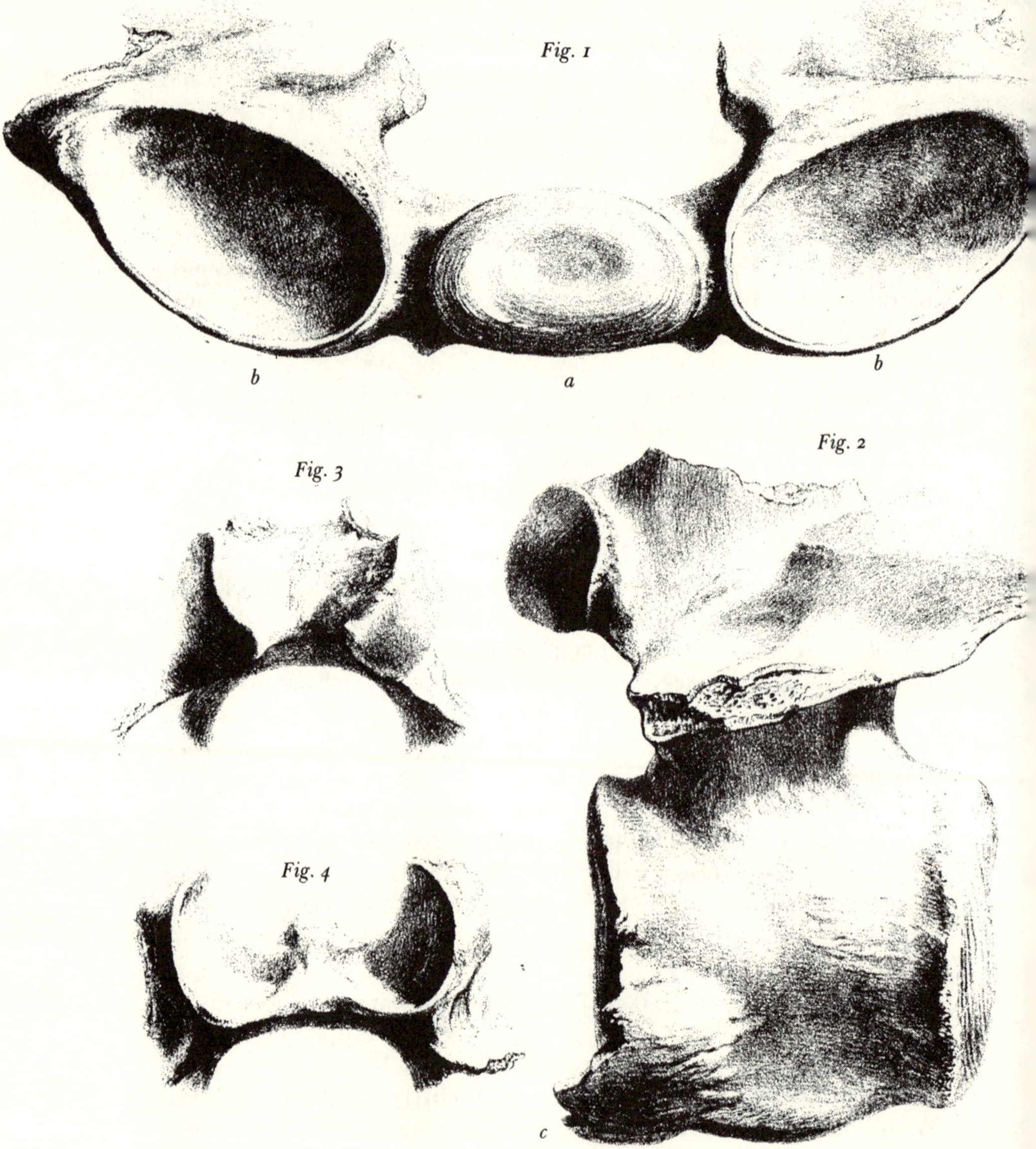

PLATE VIII

Lumbar vertebrae, Macrauchenia

Fig. 1. Posterior view of last lumbar *Figs 2, 3, 4. Fourth lumbar vertebra* smooth concavity, adapted to a corresponding convexity in the transverse processes of the first sacral vertebra. These articulating surfaces have evidently been covered with smooth cartilage; they present a pretty regular transverse ellipsoid form. A view of the three joints by which, independently of the two oblique processes, the last lumbar vertebra of the Macrauchenia was articulated with the sacrum, is given in Plate VIII, fig. 1. The transverse processes of the posterior lumbar vertebra, besides their agreement with those of the Horse and Hippopotamus in the structure just described, also correspond with them in general form, and deviate remarkably from those of the *Camelidae* in their great breadth.

It will be seen that the articulations on the body and transverse processes of the last lumbar vertebra of the Macrauchenia differ from the corresponding articular surfaces of the Horse, inasmuch as the middle surface is convex, while the two lateral ones are concave, and these are moreover relatively larger than either in the Horse or Hippopotamus: by this structure the trunk was more firmly locked to that segment of the vertebral column, which receives and transmits to the rest of the body the motive impetus derived from the hinder extremities, which are in all quadrupeds the chief powers in progression; while at the same time the shock must have been diminished by the great extent of interposed elastic cartilages; and a certain yielding or sliding motion would be allowed between the lumbar vertebrae and sacrum.

The anterior oblique processes of the lumbar vertebrae of the Macrauchenia / (fig. 4, Pl. VIII) have concave articular facets turned towards, and nearly continued into, each other at their lower extremities; so as to form together a deep semilunar notch, into which the corresponding convex articular surfaces of the posterior oblique processes of the adjoining vertebra (fig. 3, Pl. VIII) are firmly locked. In the close approximation of the two anterior concave articular facets, which are separated from each other only by a vertical ridge, and a rough surface of about three or four lines in breadth, the lumbar vertebrae of the Macrauchene resemble those of the Horse, and differ from those of the Camel-tribe and Ruminants generally, in which those surfaces are wider apart. In the hook-like form, however of these articular processes the lumbar vertebrae of the Macrauchene differ from those of the Horse; and resemble those of many Ruminant species and of the Anoplothere;[18]

[18] Cuvier, *Ossemens Fossiles*, iii, p. 238.

but the degree of concavity of the articulating surface is not so great in the Macrauchene. It would be interesting to determine the relations which the lumbar vertebrae of the Macrauchene bear to those of the Palaeothere; but the indication which Cuvier gives of the single lumbar vertebra, of which he had cognizance in the latter genus[19] is too slight to enable me to enter upon the comparison.

The whole length of the lumbar region in the Macrauchene is twenty inches. When the bodies of these vertebrae are naturally adapted together, they form a slight curve, indicating that the loins of the Macrauchene were arched, or bent downwards towards the sacrum. That the lumbar vertebrae were rigidly connected together, or but slightly flexible, is evident from the flatness of the articular surfaces of the vertebral body, and by the circumstance of ossification having extended along the anterior vertebral ligaments, and produced an anchylosis between the fourth and fifth lumbar vertebrae; (fig. 2, *c*, Pl. VIII). This kind of ossification is frequent in aged horses, and I have seen an example of a similar anchylosis of the lumbar vertebrae, by abnormal deposition of bone in their anterior ligaments, in the skeleton of a Hippopotamus preserved in the Senkenbergian Museum, at Frankfurt.

In preparing the preceding account of the cervical and lumbar regions of the vertebral column of the Macrauchene, I have felt frequently a strong desire to enter into a comparison between them and the corresponding vertebrae of the extinct Pachyderms of the Paris Basin. Some of these, as the *Anoplotherium gracile*, in the length and slenderness of the cervical vertebrae, resemble both *Auchenia* and *Macrauchenia*; others, the *Palaeotherium minus*, and probably the rest of the genus, resemble the *Camelidae* and *Macrauchenia* in having seven lumbar vertebrae. Cuvier points out the resemblance which the atlas of the Anoplothere bears to that of the Camel, and especially of the Llama;[20] but he / expressly notices the existence of the canals for the vertebral artery in the fifth or sixth cervical vertebra of the *Anoplotherium commune*.[21] Do the cervical vertebrae – say from the third to the sixth inclusive – of the *Palaeotherium* present an imperforate condition of their transverse processes, or exterior part of their sides? Cuvier, who seems not to have been aware of this peculiarity in the Camelidae, merely notices the absence of these

[19] Loc. cit., p. 234.
[20] Loc. cit., p. 235.
[21] Loc. cit., p. 237.

arterial foramina in the last cervical vertebra of the *Palaeotherium minus*,[22] which, unfortunately for the comparison I am desirous of establishing, is that which most commonly presents this imperforate condition in the Mammalia generally. As, however, the cervical vertebrae of the Palaeothere had the anterior articular surface of the body convex, and the transverse processes produced into descending laminae, it is most probable that they corresponded with the cervical vertebrae of the typical Pachyderms in the condition of their arterial foramina.

The sacrum and ossa innominata in the present specimen of *Macrauchenia* are very imperfect; but sufficient is preserved to show that the sacrum was anchylosed to the ilia: the lower boundary of this anchylosis is marked below by an external ridge, and by vascular canals and grooves in the substance of the bone, as in the Hippopotamus. The body of the sacrum is lost, but the smooth articular convexities upon the transverse processes adapted to the articular depressions of the last lumber vertebra are fortunately preserved.

The remains of the anterior extremity of our Macrauchenia include fragments of a left scapula; the proximal extremities of the anchylosed bones of the right antibrachium; the metacarpal and most of the phalangeal bones of the right fore-foot. The first-mentioned fragments, include the head and neck of the scapula, a small part of its body with the beginning of the spine, the coracoid process, and the nearly entire glenoid cavity. This articular surface (fig. 2, Pl. IX) resembles in its general form, and degree of concavity, that of the Camel and Rhinoceros, and is deeper than in the Hippopotamus. The coracoid process is represented by a slightly produced rough, thick, and obtuse tuberosity, situated closer to the glenoid cavity than in the *Camelidae* or *Rhinoceros*, and having almost the same relative position and size, as in the *Palaeotherium crassum*. The superior border or costa of the scapula presents much variety in the Ungulate quadrupeds with which we have to compare the Macrauchenia. In the Ruminants its contour forms behind the coracoid a concave sweep, which advances close to the spine of the scapula. In the Camel and Horse the marginal concavity is shallower, and the distance of the superior costa from the spine of the scapula is greater; the extent of the supra-spinal fossa increases in the true Pachyderms, and the Macrauchene agrees with them in this structure. In the Tapir,

[22] Loc. cit., p. 232.

however, / the contour of the superior costa is broken by a deep round notch immediately behind the coracoid: in the Hippopotamus this process arches in a slight degree backward over a corresponding but wider and shallower notch. In the *Palaeotherium crassum* the concavity of the superior costa, behind the coracoid, is as slight as in the Rhinoceros; but in the Macrauchenia the superior costa of the scapula begins to rise or stretch away from the parallel of the spine, immediately behind the coracoid process. The modifications of the spine of the scapula which characterize respectively the Ruminants and Pachyderms have been clearly and concisely set forth by Cuvier, who at the same time points out the exceptional condition which the *Camelidae* present in the production of the acromial angle. It was with peculiar interest and care, therefore, that I reunited all the fragments of the scapula of the Macrauchene, in the hope of gaining from this part of the skeleton as decisive evidence of an affinity to the Camel as the cervical vertebrae had afforded. It unfortunately happens, however, that the part of the scapula most important in this comparison is broken off; yet from this very circumstance, combined with a slight inclination forwards of the anterior margin of the spine immediately beneath the fractured acromion, and from the thickness of the fractured surface, we may infer that the acromial angle of the spine was more produced than in the ordinary Ruminants, although evidently in a less degree than in the Camel tribe. The Macrauchenia, however, surpasses these aberrant Ruminants, and equals the Pachyderms in the elevation and extent of its scapular spine: but this process commences about half an inch behind the glenoid cavity, and rises at once to the height of three inches above the plane of the scapula; in which structure we may trace the same tendency to the Ruminant type, as is manifested in the scapula of the Hippopotamus and Anoplotherium; for in most other Pachyderms the spine increases gradually from its extremities to the middle part. The anterior margin of the spine beneath the short acromion is perforated by an elliptical fissure measuring ten lines, by three lines. The extent of the spine which is preserved, measures eight inches and a half; it is a thin and nearly straight plate of bone, expanding into a thick and rugged upper margin, which slightly over-arches the inferior fossa (fig. 1, Pl. IX). In its general form and proportions the spine of the scapula in Macrauchenia presents the nearest resemblance to that of the Hippopotamus; but its origin is closer to the articular surface of the scapula than in this, or any other Pachydermal or Ruminant genus.

Fig. 1

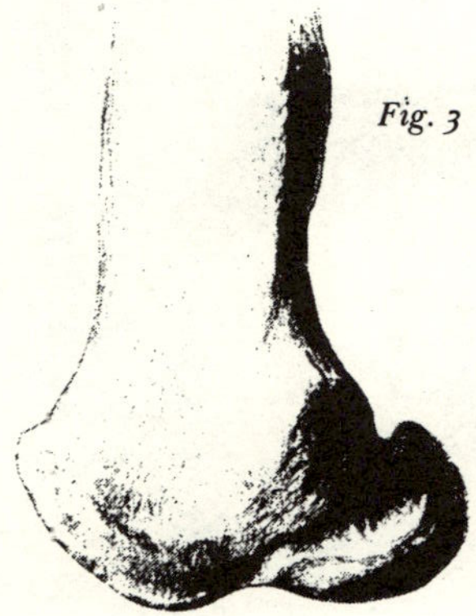
Fig. 3

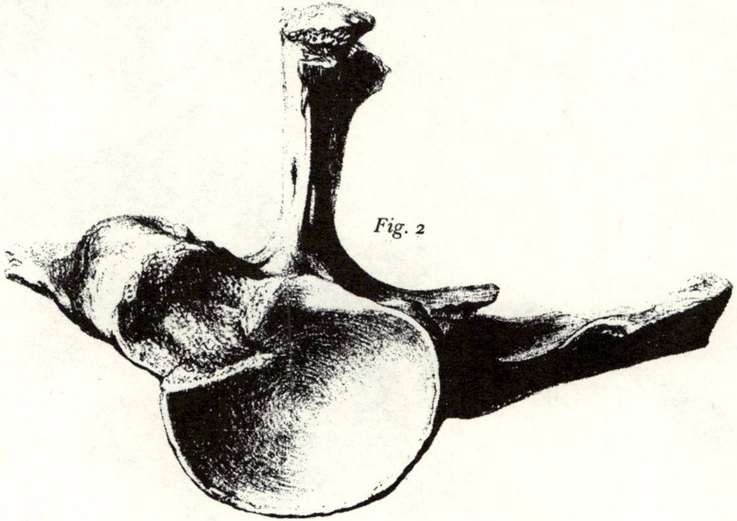
Fig. 2

PLATE IX

Macrauchenia

Figs 1, 2. Scapula *Fig. 3. Femur, drawn ½ the size of Figs 1 and 2*

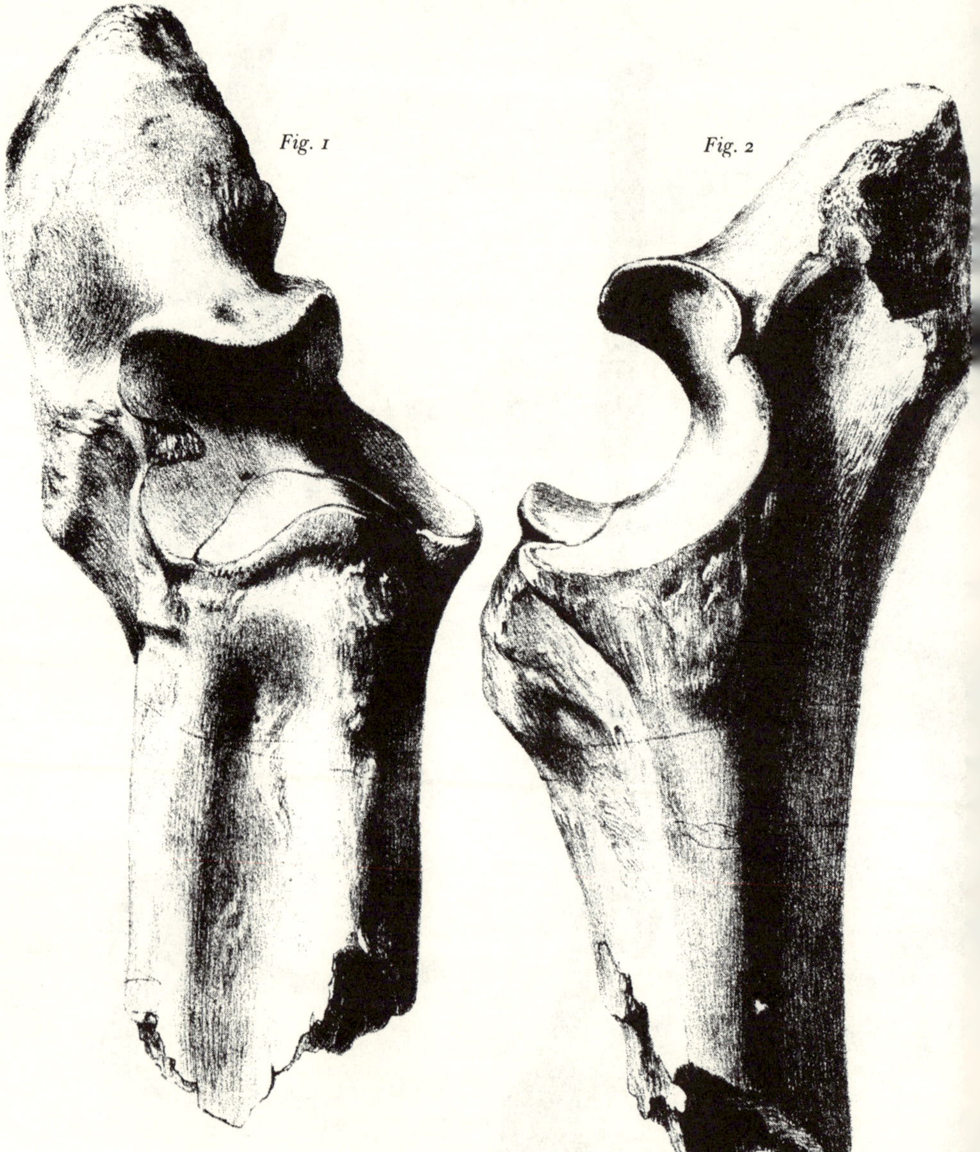

PLATE X

Proximal extremity of anchylosed ulna and radius, Macrauchenia

The portion of the antibrachium of the Macrauchenia which is preserved, presents a condition of the radius and ulna intermediate to those which respectively characterize the same bones in the Pachyderms and Camels. In the former the radius and ulna are separate bones, united in the prone position by ligament, yet so that the movement of supination cannot be performed; in the / ordinary Ruminants they are partially joined by bony confluence, which rarely extends to the proximal extremities; in the Camel and Llama the anchylosis of the radius and ulna is so complete, that no trace of their original separation can be perceived, and the olecranon appears but as a mere process of the radius.

In the Macrauchenia the anchylosis of the radius and ulna is also complete, but the boundary line of the two originally distinct bones is very manifest, and the proportion which each contributes to the great articulating surface for the distal end of the humerus is readily distinguishable. About a sixth part of this surface is due to the head of the radius, which enters into the composition of the anterior and outer part of the articulation, and its extent is defined by a depressed line describing a pretty regular curve, with the concavity directed forwards and a little outwards (*a*, fig. 1, Pl. X). Just below the articular surface a strong triangular rugged protuberance projects from the front of the head of the radius, for the attachment of the tendon of the biceps. The line of separation of the radius and ulna is indicated on the inner side of the head of the radius by a deep and narrow fissure extending downwards from below the anterior part of the articulating surface; and on the outer side by a broad groove leading upwards to a deep pit near the proximal end of the antibrachium. We may see by the direction of the head of the radius which is thus defined, that it crosses obliquely in front of the ulna, as in the Elephant, Hippopotamus, and other Pachyderms, and that the bones are anchylosed in the prone condition: below this fissure and groove, which mark the interosseous line, the radius and ulna become blended together into one compact bone, which is flattened from before backwards, with a well marked ridge on the outer side; and excavated by a single medullary cavity, the compact walls of which present a general thickness of one-third of an inch.

The proximal articular surface of sigmoid cavity of the antibrachium, constituted as above described, resembles that of the Palaeothere, Tapir, and the generality of the Pachyderms in having two depressions, instead of three, as in the Anoplothere, and

Ruminants. The Hippopotamus has a slight tendency to the latter structure, which is also less marked in the Camel than in the ordinary Ruminants. In its general form the sigmoid cavity of the Macrauchene resembles that of the Hippopotamus more than that of the Camel. In the Camel this articular surface is traversed transversely by a broad, shallow, and slightly roughened tract, which divides the smooth surface of the joint into two parts, one forming the anterior horizontal surface due to the conjoined radius and ulna, the other forming the vertical concave surface on the anterior part of the base of the olecranon. In the Hippopotamus there is, as it were, an attempt at a similar division of the articulating surface at the proximal end of the antibrachial bones; a deeper and rougher depression encroaches upon the articulation from its outer side, but stops when it has reached half-way across. In the / Macrauchenia the roughened surface, (*b*, fig. 1, Pl. X) commencing also at the outside, extends only one-third of the way across the articular surface: it is, however, as shallow as in the Camel. The articular surface on the anterior part of the base of the olecranon is broader in the Hippopotamus than in the Camel; but in the Macrauchene it is twice as broad as in the Hippopotamus. The size of the olecranon in the Macrauchene exceeds that of the Hippopotamus, and *à fortiori* that of the Camel: indeed in its general magnitude the Macrauchenia must have fully equalled the largest Hippopotamus; but it no doubt had a more shapely, and less broad and bulky trunk. The olecranon of the Macrauchenia differs in shape, both from that of the Camel and Hippopotamus; it terminates above in a three-sided cone with an obtuse apex; and presents a well marked protuberance at the outer side of the base, which is not present in either the Camel or Hippopotamus. There is also a strong rugged ridge on the back part of the olecranon which makes an angle before sinking into the level of the ulna below.

The confirmation of the close affinity of the Macrauchenia to the Pachydermatous Order, which the structure of the cervical vertebrae alone might have rendered very doubtful, is afforded by the bones of the right fore-foot (Pl. XI); these are fortunately in so perfect a condition, as to make it certain that this interesting quadruped had three toes on the fore-feet, and not more; and that the fully developed metacarpal bones are distinct, and correspond in number with the toes, and are not anchylosed into a single cannon bone, as in the Ruminants. The bones preserved are the metacarpals, proximal phalanges, and middle phalanges of each of the three toes, and the distal phalanx of the innermost toe.

The proximal end of the innermost metacarpal bone presents three articular surfaces; the middle facet is the largest, and the two lateral ones slope away from it at an angle of 45°. The middle facet is broad and slightly convex in front, narrow and concave behind; the distal articular surface of the trapezoides must have corresponded with this surface; the outer facet is narrow, flat, extends from the fore to the back part of the head of the bone, and must have been adapted to a corresponding surface on the os magnum; the inner facet is the smallest, presents a triangular form, and is situated towards the back part of the head of the metacarpal bone; it indicates the existence of a rudimental metacarpal bone, or vestige of a pollex. Below the outermost of the lateral surfaces there is a crescentic articular surface with its concavity directed outwards and downwards (fig. 2, Pl. XV), against which a corresponding convex articular surface of the middle metacarpal abuts (fig. 3, Pl. XV). External to this surface the proximal end of the middle metacarpal bone presents two articular surfaces for the carpus; the larger one, which was adapted to the os magnum, is horizontal, / broad and convex before, narrow and concave behind; the outermost facet is a small triangular surface inclined downwards to the level of the articulating surface of the outermost metacarpal. It also presents a posterior vertical articular surface for a sesamoid bone. The proximal extremity of the outer metacarpal bone is joined to the middle metacarpal, not by one semilunar surface, but by two separate articulations of small size (figs 4 and 5, Pl. XV); it presents a single large slightly convex articular surface for the os magnum, of an irregular semicircular form, with the convexity of the curve turned outwards.

The metacarpus increases in breadth as it approaches the phalanges; the two lateral metacarpals bending slightly away from the middle one, and expanding towards their distal extremities: the middle bone presents a symmetrical figure except at its proximal extremity (fig. 2, Pl. XI). The distal articulating facet of each of the metacarpal bones extends so far upon both the anterior and posterior surfaces as to describe more than a semi-circle (fig. 3, Pl. XI); in the two lateral metacarpals it is traversed throughout by a longitudinal convex ridge dividing it into two equal lateral parts; the ridge is most produced on the posterior half of the joint (fig. 4, Pl. XI): in the middle metacarpal this ridge subsides before it reaches the anterior part of the articular surface.

The proximal extremity of the middle proximal phalanx presents a posterior notch corresponding to the above partially developed ridge: the proximal extremities of the lateral phalanges are traversed by a middle longitudinal depression, and two lateral shallow concavities (fig. 6, Pl. XI); but these are of such an extent as to be in contact with only a part of the convexity above, which therefore was doubtless adapted to a sesamoid bone on each side of the longitudinal ridge. The structure of the above described joints proves that the motion of the toe upon the metacarpus was much freer and more extensive than in the Rhinoceros, which is the only existing Ungulate mammal which presents the tridactyle structure in the fore-foot. In this species the metacarpo-phalangeal articulations exhibit only a slight trace of the longitudinal ridges and grooves which are confined to the posterior part of the joint; these are more developed in the *Camelidae*; but the Hog and Horse in this respect approach nearer to the Macrauchene, though the structure of the metacarpo-phalangeal joints in the Hog falls far short of the compactness and strength combined with freedom of play in flexion and extension which distinguish those of the Macrauchene. The *Palaeotherium medium* most resembles the Macrauchene in the structure of the trochlear metacarpo-phalangeal joints; but both in this species,[23] and the *Pal. crassum*[24] the articular surface at the distal end of the metacarpal is relatively narrower than in the Macrauchenia; moreover all the species of the extinct Palaeothere differ from the Macrauchene / in the greater size and strength of the middle as compared with the lateral metacarpals.

The articulation at the distal extremity of the proximal phalanges (fig. 5, Pl. XI) is simple, and not divided into two pulleys by a longitudinal ridge; it is slightly concave from side to side; but in its extent upon the anterior and posterior surfaces of the bone indicates a freedom of flexion and extension of the toes, which harmonizes with the structure of the joint above.

The proximal articulating surfaces of the second phalanges (fig. 7, Pl. XI) corresponds of course to those to which they are adapted; they are, however, characterized by sending upwards an obtuse process from the middle of their anterior margin. The distal articulating surfaces (fig. 8, Pl. XI) resemble those of the proximal phalanges, but extend further upon the back part of the phalanx than the front, indicating the more horizontal position of the second phalanges.

[23] See *Ossem. Fossiles*, Pl. XX, fig. 3.
[24] Loc. cit., Pl. XXII, fig. 6.

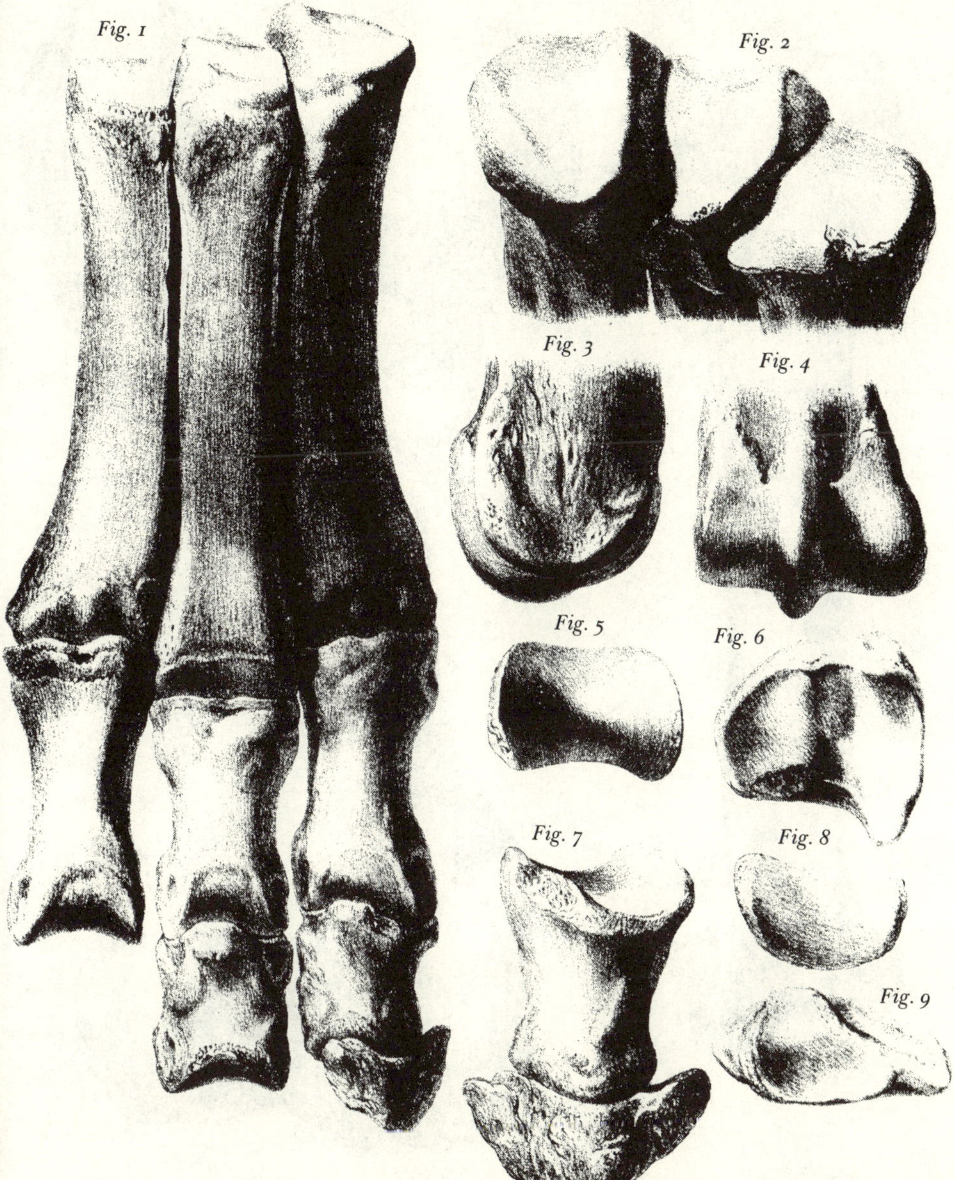

PLATE XI

Bones of the right fore-foot, Macrauchenia

Fig. 1 is drawn ⅔ the size of Figs 2–9

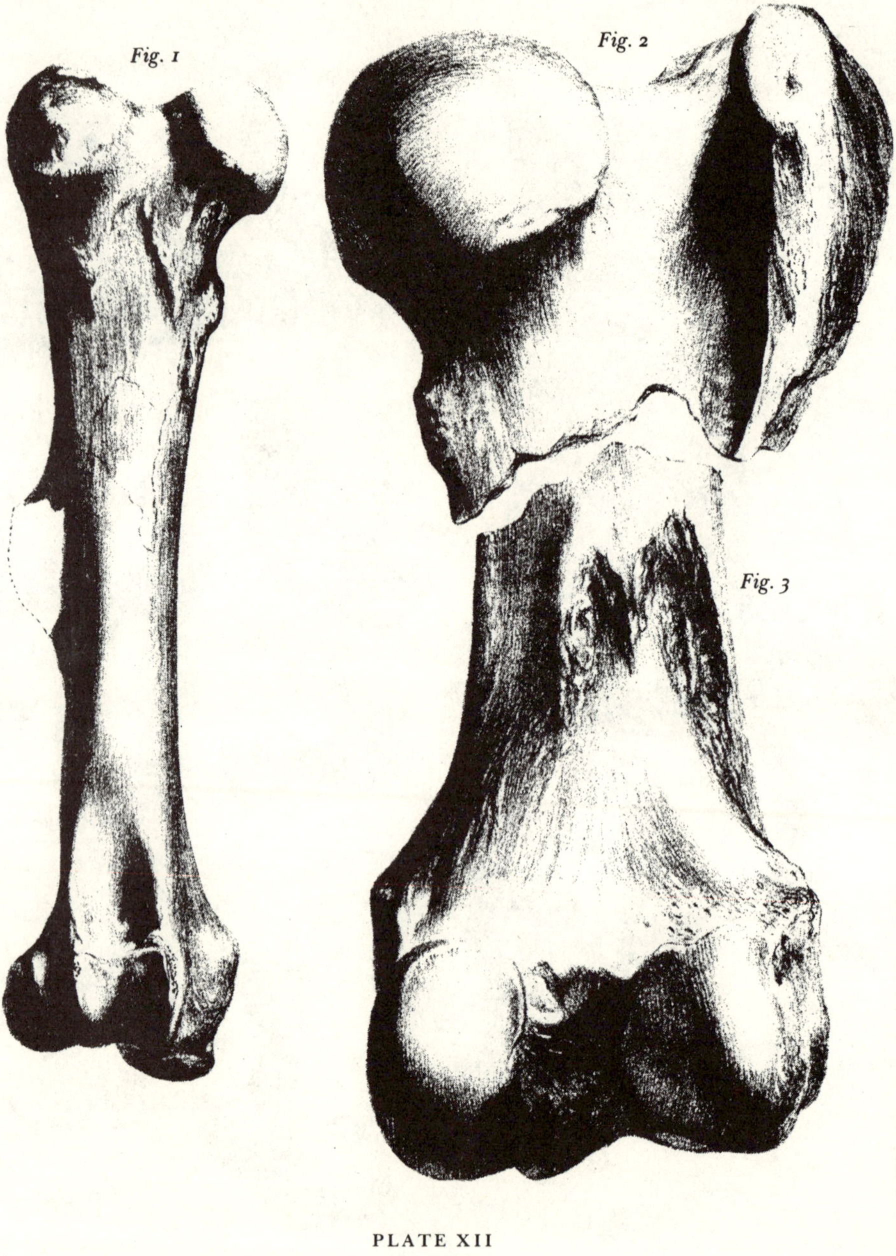

PLATE XII

Right femur, Macrauchenia

Fig. 1 is drawn ½ the size of Fig. 2.

The last phalanx, does not resemble the neatly defined ungulate phalanges of the Ruminantia, and Solipedia, but has the irregular form characteristic of those of the Pachydermata. It is wedge-shaped, broader than it is long, with a rugged surface, except where it plays upon the distal end of the second phalanx, where it is slightly concave in one direction, and convex in the other (figs 7 and 9, Pl. XI). A portion of this phalanx extends backwards behind the articular surface, as in the corresponding bone of the Palaeothere and Rhinoceros.

The femur of the Macrauchenia (fig. 1, Pl. XII) is full two feet in length, and consequently longer than in any known Camel or Rhinoceros; as compared with its transverse diameter it is much longer than the femur of the latter animal: in the proportion of its breadth to its length, and the expansion of its extremities as compared with the diameter of the shaft, it more resembles that of the Camel. The femur of the Giraffe deviates from that of the Macrauchenia in the excessive expansion of its distal extremity. But the most striking evidence deducible from this bone, of the affinity of the Macrauchenia to the true Pachydermatous type is afforded by the evident traces of a third trochanter, the outline of which is conjecturally restored in the figure. Of the Pachyderms which have this characteristic structure, the extinct Palaeothere offers the nearest resemblance to the Macrauchene in the general form and structure of the femur.

The head of the femur in the Macrauchene (fig. 2, Pl. XII) presents the form of a pretty regular hemisphere; it is less flattened above, and is directed more obliquely inwards than in the Palaeothere: the neck supporting it does not project so far from the shaft as in the Palaeothere or Tapir, but farther than in the Camel. The great trochanter rises above the level of the head; in which structure and in the depression between the head and trochanter, the femur of the Macrauchene offers a character intermediate between the Tapir or Palaeothere, and the Camel. The lesser trochanter is a slight projection from a ridge / of bone which is continued from the under part of the head of the femur to the inner surface of the shaft. In the Palaeothere the lesser trochanter is situated more towards the posterior surface of the femur; so that, in this particular, the Macrauchene approaches nearer to the Camel. Cuvier makes no mention of the condition of the depression for the *ligamentum teres* in the Palaeothere. Among existing ordinary Pachyderms the Hippopotamus presents no trace of the insertion of a *ligamentum teres* in the head of the femur; in the Camel the place of its insertion is indicated by a well-marked circumscribed pit; in the Tapir a similar circular depression is situated close to the inferior margin of the articular convexity. The ligament was undoubtedly present in Macrauchenia, but the place of its insertion is a broad and deep notch leading from the under and back part of the head of the bone a little way into its articular surface: this I regard as another of those interesting transitional structures with which the remains of the Macrauchenia, few and imperfect though they unfortunately are, so freely abound.

The femur of Macrauchenia, in the flatness of the back part of its neck, and the elongated form of the post-trochanterian depression, resembles that of the Camel rather than that of the Palaeothere; and the same resemblance is shown in the cylindrical figure, straightness, and length of the shaft. The depth of the trochanterian depression, and the incurvation of the strong ridge continued downwards from the great trochanter are individual peculiarities in the Macrauchenia.

A great part of the third trochanter is broken off; but from the remains of its base we see that it had the same relative size as in the Palaeothere; but it is situated at the middle of the shaft of the femur, and consequently lower down than in the Palaeotheres and Tapirs. In the general form and relative size of the condyles at the distal extremity of the femur (fig. 3, Pl. IX and XII) the Macrauchene is intermediate to the Camel and Palaeothere, but resembles more the latter. In the articular surface for the patella, it deviates somewhat from the Palaeothere, having this part longer in proportion to its breadth, more regularly and deeply concave from side to side, and with its lateral boundaries more sharply defined. In all these points the Macrauchene approaches the Camel: the same affinity is shown in the protuberance above the inner condyle; but in the extent of the posterior projection of this condyle (fig. 3, Pl. IX) it exceeds the Camel and Palaeothere, and displays an intermediate structure between these species and the Hippopotamus.

There is a rough crescentic depression above the outer condyle where the linea aspera begins to diverge; the corresponding depression is deeper in the Hippopotamus, while in the Camel it is represented by a roughened surface only, which is not depressed. In the fossa between the rotular articulation and the external condyle the Macrauchene resembles the Camel: the interspace of the / condyles is relatively wider than in the Camel, and the process above the inner condyle is more angular; in both these respects the Macrauchene inclines towards the Palaeothere.

In the structure of the bones of the leg of the Macrauchenia we find the same transitional character which is afforded by the definable limits of the anchylosed bones of the fore-arm. In the Pachyderma the fibula is an entire and distinct bone. In the Ruminantia, with the exception of the small Musk-deer, and, in an inferior degree, the Elk, the fibula appears only as a short continuous process sent down from the under part of the external condyle of the tibia. In the Camel tribe the only trace of the fibula in the bones of the leg, is this process in a still more rudimental state. In the Macrauchenia the fibula is entire, but is confluent with the tibia through nearly its whole extent: the proximal part of the fibula is well defined; its head is anchylosed to the outer condyle of the tibia, but the shaft is continued free for the extent of nearly two inches, and then again becomes confluent with the tibia, forming apparently the outer ridge of that bone. About five inches from the distal end of the tibia this outer ridge becomes flattened by being, as it were, pressed against the tibia, and the anterior and posterior edges are raised above the level of the tibia; beyond this part the limits of the fibula begin again to be defined by deep vascular grooves. The outer side of the distal end of the fibula is excavated by a broad tendinous groove. The fibula and tibia are distinct bones in both the Palaeothere and Anoplothere, as in the Pachyderms. It is to the former genus, however, especially *Pal. magnum*, that the Macrauchene presents the nearest approach in the general form of the tibia, the principal bone of its leg: but in the Macrauchene the tibia is relatively shorter, and thicker, and is straighter and less expanded at its extremities, especially the upper one, than in any of the Palaeotheres.

The mesial boundaries of the two superior articulating surfaces of the tibia are raised in the form of ridges, which are separated by a deep groove; of these ridges the external is the highest, as in *Pal. magnum*: but the articular surfaces in the Macrauchene slope away from these ridges more than in the Palaeotheres. The rotular or anterior tuberosity of the tibia is more produced, and rises higher than in the Palaeotheres; the ridge continued downwards from this process is more marked in the Macrauchene, and its limits are better defined: the shaft of the tibia below the ridge is also more flattened in the antero-posterior direction than in the Palaeothere. The configuration of the back part of both proximal and distal extremities of the tibia are so clearly and accurately given in figures 2 and 3, Pl. XIII, as to render verbal description unnecessary. Neither the text nor the figures in the *Ossemens Fossiles* afford the means of pursuing the comparison

Fig. 1

Fig. 2

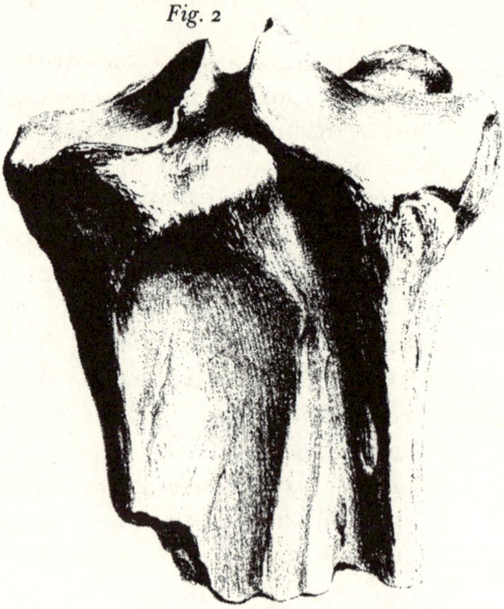

Fig. 3

Fig. 4

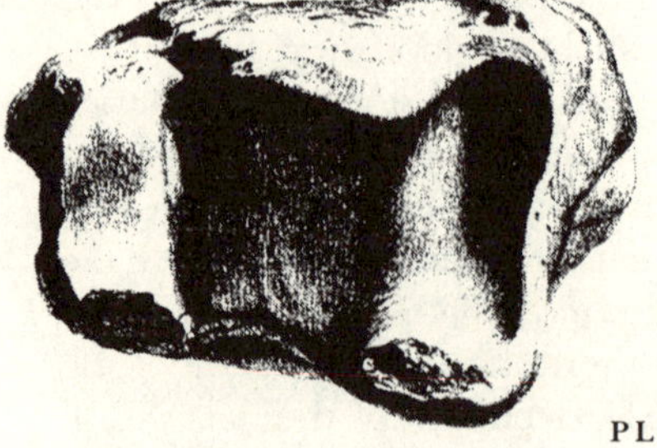

PLATE XIII

Right tibia and fibula, Macrauchenia

Fig. 1 is drawn ½ the size of Figs 2–4.

between the Macrauchene and Palaeothere in these particulars; and I proceed, / therefore, to the consideration of the inferior articulating surface of the bones of the leg (fig. 4, Pl. XIII).

Since, of the hind foot, we possess in the present collection only a single tarsal and metatarsal bone, the structure of the distal articular surface of the tibia is attended with peculiar interest, because we are taught by Cuvier that it reveals to us in the Ungulate animals the didactyle or tridactyle structure of the foot. In the Ruminants this articular surface is nearly square, and extended transversely between two perpendicular malleoli, while in the Pachyderms with three toes to the hind-foot the articular surface of the tibia is oblique, and is divided into two facets between the perpendicular malleolar boundaries. Now in the Macrauchenia, although the two bones of the leg are anchylosed together, the extent of that part of the tarsal articular surface which is due to the tibia is indicated, as in the case of the radius in the joint of the fore-arm, by a groove; and we are thus able to compare this surface with the distal articular surface of the tibia in the Palaeothere and Anoplothere. It presents in the Macrauchenia a very close resemblance with that of the *Palaeotherium magnum*,[25] being divided into two facets by a convex rising, which traverses the joint from behind forwards; but the ridge is narrower, the internal facet somewhat deeper, and the external oblique surface rather flatter than in the three-toed Palaeothere. In the portion of the tarsal articular surface due to the fibula, we find, however, a more marked deviation from the Palaeothere, and an interesting correspondence with the Anoplothere, in the inferior truncation and horizontal articular surface which is continued upon the lower extremity of the fibula, at right angles with the vertical malleolar facet which forms the outer boundary of the trochlea of the astragalus: this articular surface unerringly indicates a corresponding articular projection in the calcaneum, which, therefore, although the bone itself does not form part of the present collection, we may conclude to differ from the calcaneum of the Palaeotherium, and to resemble that of the Anoplotherium, in this particular at least.

The valuable indication which the distal articular surfaces of the anchylosed tibia and fibula have given of the correspondence of the hind-foot with the fore-foot of the Macrauchenia, in regard to the number of the toes, receives ample confirmation from the astragalus, which, of all the bones in the foot, is the one that an anatomist would

[25] See *Ossem. Foss.*, iii, Pl. XXVI, fig. 5.

have chosen, had his choice been so limited, and which most fortunately has been secured by Mr Darwin, in a very perfect state, in the present instance. I have compared this astragalus with that of the Giraffe, and other Ruminants, the Camel, the Anoplothere, the Horse, the Hog, the Hippopotamus, Rhinoceros, Tapir, and Palaeothere: it is with the Pachyderms having three toes to the hind-foot, that the Macrauchenia agrees in the main distinguishing / characters of this bone; its anterior articular surface, for example, is simple, and not divided into a double trochlea by a vertical ridge: lastly, it is with the astragalus of the Tapir and Palaeothere that it presents the closest correspondence in the general form and the minor details of structure, and with these Pachyderms, therefore, I shall chiefly limit the comparison of the Macrauchenia, in regard to the bone in question. If the upper or tibial articular surface (fig. 5, Pl. XIV) be compared with that in the *Palaeotherium magnum* (*Ossem. Foss.*, Pl. LIV, fig. 2), it will be seen, that the general direction of that surface is more parallel with the axis of the bone in Macrauchenia. In the Palaeotherium it is turned a little towards the outer or fibular side, and in the Tapir the general direction of the same surface is placed still more obliquely. The anterior border of this articulating surface is broken by a semicircular notch in the Palaeothere; in the Tapir it describes a gentle concave curve, and the Macrauchene resembles the Tapir in this respect. The chief difference between the astragalus of the Tapir and the Palaeothere, when viewed from above, obtains in the relative length of the bone, anterior to the tibial articulating surface: the Macrauchene presents, in this respect, an intermediate structure, but differs from both in the greater extent of the tibial side of this part of the astragalus.

If we next direct attention to the anterior or scaphoid articular surface (fig. 3, Pl. XIV), and compare it with that of the *Palaeotherium magnum* (fig. 4, Pl. liv, *Ossem. Foss.*), it will be seen, that it presents in the Macrauchenia an oval, and in the Palaeotherium an irregular quadrangular form: in the Macrauchenia, this surface is uniform or undivided, and is gently convex, except at its lower part; while in the Palaeothere it is divided by an oblique ridge into a broad internal facet for the scaphoid bone, and a narrow internal surface for articulation with the os cuboides; the larger surface is also concave transversely, and slightly convex vertically: in the Tapir, the anterior surface of the astragalus deviates still further from that of the Macrauchenia, both in general form, and in the proportion of the cuboidal facet. In the

Fig. 1

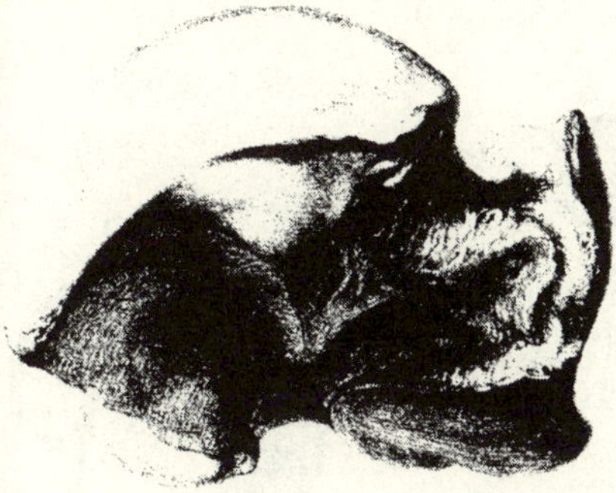

Fig. 2

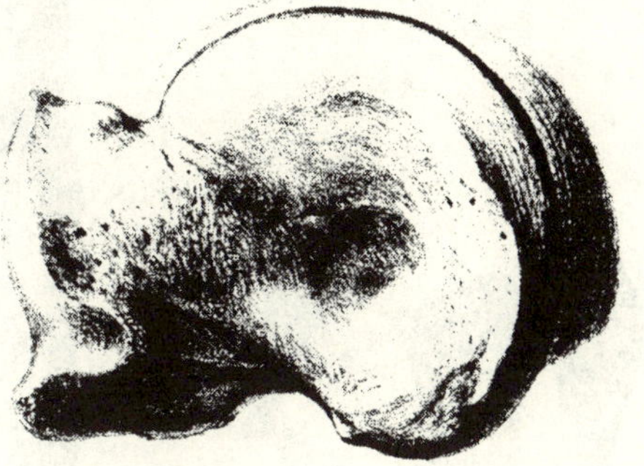

Fig. 3

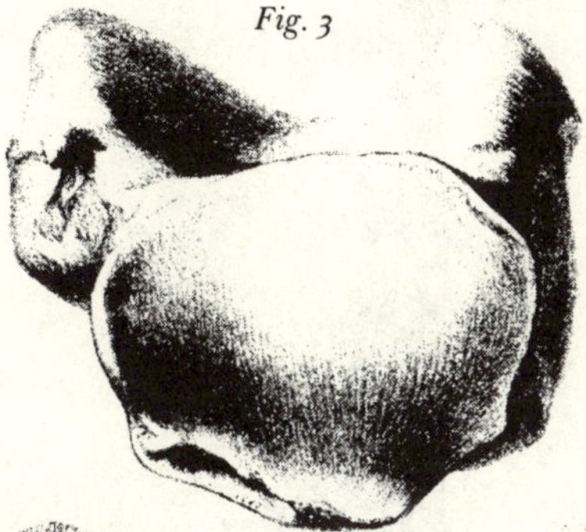

Fig. 4

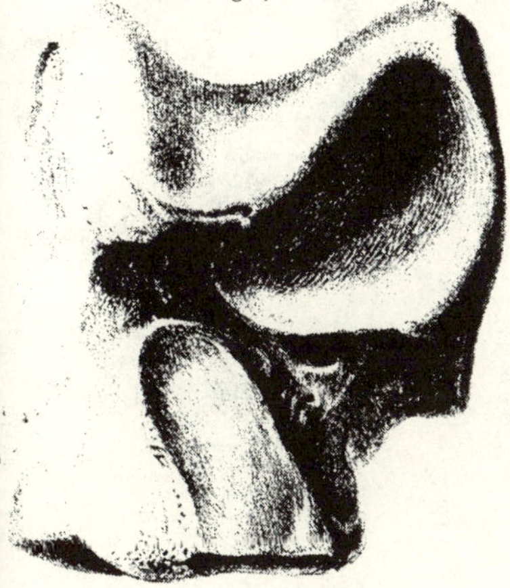

Fig. 5

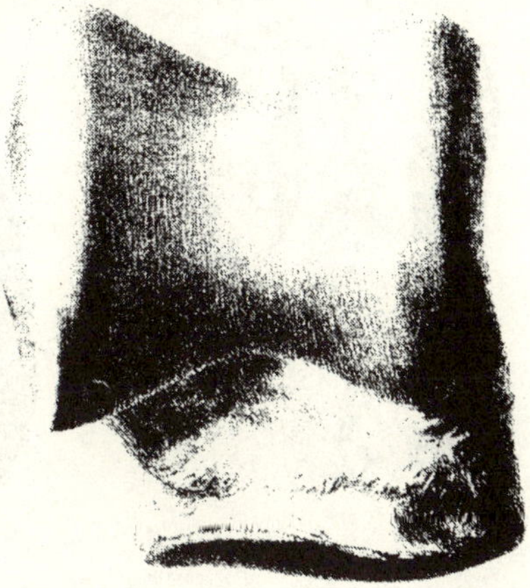

PLATE XIV

Right astragalus, Macrauchenia

Fig. 1

Fig. 2

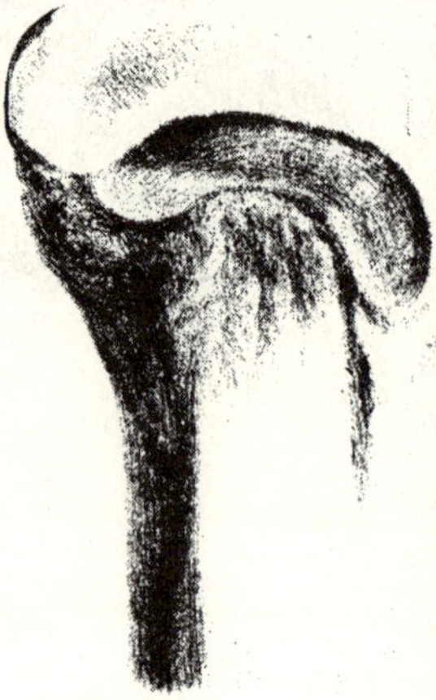

Fig. 3

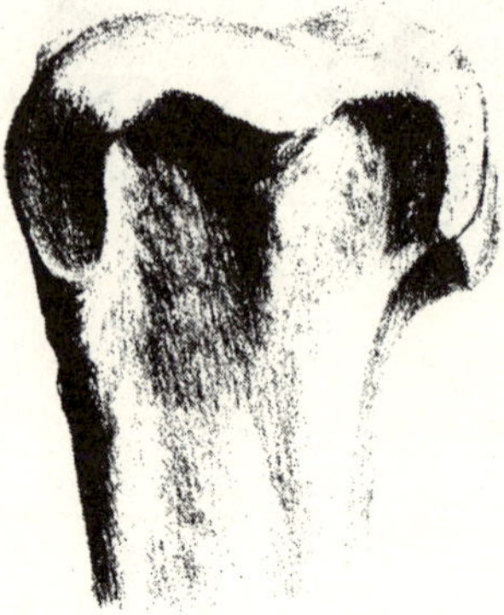

Fig. 4

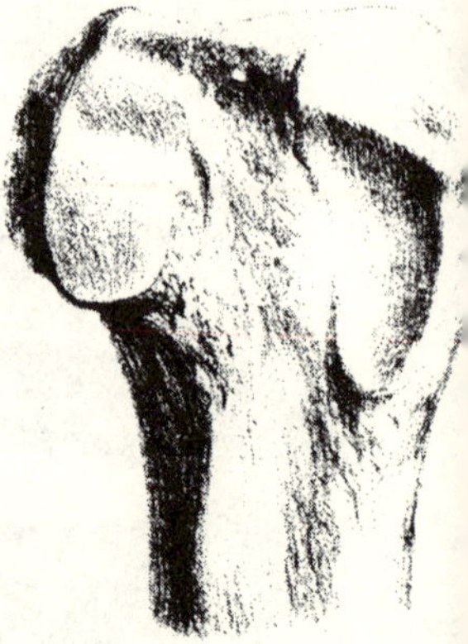

Fig. 5

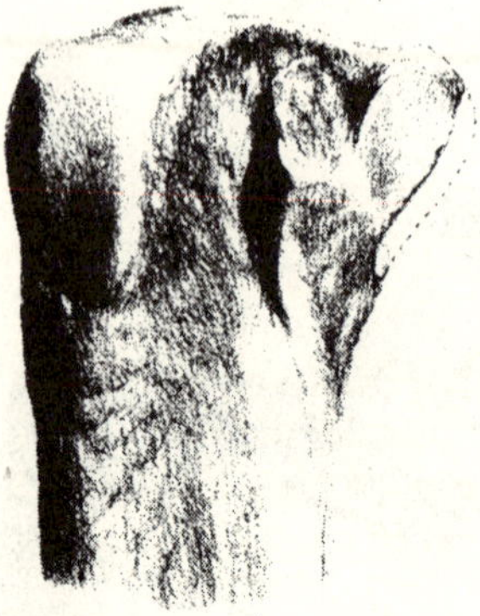

PLATE XV

Macrauchenia

Fig. 1 Metatarsal *Figs 2–5 Metacarpals*

didactyle Anoplotherium, Camel, and true Ruminants, where the cuboides presents a large relative size, a still greater proportion of the anterior surface of the astragalus is devoted to the articulation with this bone, and is separated from the scaphoid surface by a well-developed vertical ridge. The Macrauchenia presents, therefore, the extreme variation from this type; and should the entire tarsus hereafter be discovered, it will doubtless be found, that the os cuboides is articulated posteriorly to the os calcis exclusively.

The external surface of the astragalus of the Macrauchene (fig. 1, Pl. XIV), is longer in proportion to its vertical extent than in the Tapir or Palaeothere: the articular surface for the fibular malleolus is less curved. Between this surface and the anterior facet the bone is excavated by a deep notch, both in the Tapir and Palaeothere; but in the Macrauchenia by a gentle concavity. Beneath the / malleolar articular smooth surface in the Palaeothere there is a deep pit; in the Tapir a shallow one; but in the Macrauchenia we observe only a smooth and slightly convex triangular surface. If we compare the inner surface of the astragalus in these three animals, we shall find the existing Tapir again forming a transition between the two extinct genera. In the Palaeothere, a round protuberance projects from the anterior part of this surface: in the Tapir, we observe a gentle rising of the bone in the same part, while in the Macrauchene (fig. 2) the surface of the bone is level at this part. The margin of the tibial malleolar articular surface, which is very slightly raised in the Macrauchene, is more developed in the Tapir, and still more so in the Palaeothere, where it forms a ridge, overhanging the rough outer side of the bone. Near the lower part of this surface we observe a small but deep depression in the Palaeothere; there is a shallower one in the corresponding part in the Tapir; and the depression is still wider and shallower in the Macrauchenia. In the Palaeothere the astragalus articulates by three surfaces with os calcis, posteriorly by a large concave surface, externally by a longitudinal sub-elliptic surface, and anteriorly by a thin transverse facet: in the Macrauchene (fig. 4) two only of these surfaces are present, viz. the concave and the longitudinal one, the anterior transverse surface being wanting: in the Tapir, the transverse surface is present, but is confluent with the longitudinal one. The posterior surface is relatively larger and deeper in the Macrauchene than in the Palaeothere, and approaches nearer to the triangular than the oval form: the longitudinal surface is placed more obliquely, and is truncated anteriorly. In the Tapir this surface is confluent with the scaphoid articular surface, but it is separated therefrom by a narrow strip of bone in both the Palaeothere and Macrauchene. It is satisfactory to find in the bone, which marks most strongly the affinity of *Macrauchenia* to *Palaeotherium*, so many easily recognizable differences, because the structure of the cervical vertebrae in the latter genus is too imperfectly known, to allow us to predicate confidently a distinction between it and *Macrauchenia* in that particular; the difference, however, which they present in the condition of the bones of the fore-arm and leg, forbids their being considered as generically related.

There remains to be noticed only a single fractured metatarsal bone (fig. 1, Pl. XV). This, from its bent and unsymmetrical figure, is evidently not a middle one, and having the side of the proximal end, which was articulated to the adjoining metatarsal in a nearly perfect state, it enables us to refer it with certainty to the hind-foot, since it does not agree with any of the corresponding surfaces at the proximal extremities of the metacarpal bones. It remains then to be determined, whether it is an external metatarsal of the right-foot, or an internal one of the left-foot, the general curvature of these being in the same direction. With neither of these bones in the Tapir does our metatarsal agree, since it has but one articular facet on the lateral / surface of its proximal end, while the outer metatarsal of the right-foot of the Tapir, with which, in other respects, it most closely corresponds, has two articular surfaces. In the cast of a hind-foot of a Palaeothere, I find that the outer metatarsal bone closely agrees with this metatarsal bone of the Macrauchene, in the structure just alluded to: the articulation with the middle metatarsal being by a single sub-oval facet, which stands out a little way from the surface of the bone: the articular surface in the Macrauchene presents a similar form and condition, and is similarly situated to that in the Palaeothere, being at the posterior part of the lateral surface, and a little below the superior or tarsal articular surface. The bone expands towards its distal end, which corresponds in structure with those of the two lateral metatarsals in the fore-foot, in being completely divided into two trochlear surfaces by a well-developed median ridge, and in having the posterior half of this ridge suddenly produced, so as to project about two lines further from the trochlear surface than the anterior part of the same ridge. In both the Tapir and Palaeothere this anterior part of the ridge is wholly suppressed, and the posterior is much more feebly developed than in the Macrauchenia. The metatarsal bone here described is of exactly the same length with the internal metacarpal bone, and proves, in conjunction with the proportions of the astragalus, that the fore and hind feet of the Macrauchenia were of equal size.

Thus then we obtain evidence, from a few mutilated bones of the trunk and extremities of a single representative of its race, that there once existed in South America a Pachydermatous quadruped, not proboscidian, which equalled in stature the Rhinoceroses and Hippopotamuses of the old world. But this, though an interesting and hitherto unsuspected fact, is far from being the sum of the information which is yielded by these fossils. We have seen that the single ungueal phalanx bespeaks a quadruped of the great series of the *Ungulata*, and this indication is corroborated by the condition of the radius and ulna, which are fixed immoveably in the prone position. Now in the Ungulated series there are but two known genera – the Rhinoceros and Palaeotherium – which, like the quadruped in question, have only three toes on the fore-foot. Again, in referring the Macrauchenia to the Tridactyle family of Pachyderms, we find, towards the close of our analysis, and by a detailed comparison of individual bones, that the Macrauchenia has the closest affinity to the Palaeotherium.

But the Palaeotherium, like the Rhinoceros and Tapir, has the ulna distinct from the radius, and the fibula from the tibia; so that even if the Parisian Pachyderm had actually presented the same peculiarities of the cervical vertebrae as the Patagonian one, it would have been hazardous, to say the least, while ignorant of the dentition of the latter, to refer it to the genus *Palaeotherium*.

Most interesting, indeed will be the knowledge, whenever the means of obtaining it may arrive, of the structure of the skull and teeth in the Macrauchenia. / Meanwhile, we cannot but recognize, in the anchylosed and confluent state of the bones of the fore-arm and leg, a marked tendency in it towards the Ruminant Order, and the singular modifications of the cervical vertebrae have enabled us to point out the precise family of that order, with which the Macrauchenia is more immediately allied.

In first demonstrating this relationship, it was shown in how many particulars the *Camelidae*, without losing the essential characters of Ruminantia, manifested a tendency to the Pachydermatous type; and the evidence which the lost genera, *Macrauchenia* and *Anoplotherium*, bear to a reciprocal transition from the Pachyderms to the Ruminants, through the *Camelidae*, cannot but be viewed with extreme interest by the Zoologist engaged in the study of the natural affinities of the Animal Kingdom.

The Macrauchenia is not less valuable to the Geologist, in reference to the geographical distribution of animal forms. It is well known how unlooked-for and unlikely was the announcement of the existence of an extinct quadruped entombed in the Paris Basin, whose closest affinities were to a genus (*Tapirus*), at that time, regarded as exclusively South American. Still greater surprise was excited when a species of the genus *Didelphys* was discovered to have co-existed in Europe with the *Palaeotherium*.

Now, on the other hand, we find in South America, besides the Tapir, which is closely allied to the Palaeothere – and the Llama, to which the Anoplothere offers many traces of affinity – the remains of an extinct Pachyderm, nearly akin to the European genus *Palaeotherium*: and, lastly, this Macrauchenia is itself in a remarkable degree a transitional form, and manifests characters which connect it both with the Tapir and the Llama.

ADMEASUREMENTS OF THE BONES OF THE MACRAUCHENIA

	Inches	*Lines*
Length of third (?) cervical vertebra	7	9
Vertical diameter of cervical vertebra	4	0
Vertical diameter of body of cervical vertebra	2	3
Transverse diameter of cervical vertebra	3	3
Vertical diameter of spinal canal	1	–
Length of fourth lumbar vertebra	5	5
Vertical diameter of body of lumbar vertebra	2	9
Transverse diameter of lumbar vertebra	2	10
Vertical diameter of spinal canal	1	1
Transverse diameter of spinal canal[26] /	1	6
Transverse diameter of last lumbar vertebra	9	–
Transverse diameter of body of lumbar vertebra	2	2
Vertical diameter of lumbar vertebra	1	3
Entire length of lumbar region of vertebral column	20	–
Vertical diameter of glenoid cavity of scapula	3	–
Transverse diameter of glenoid cavity of scapula	2	10
Elevation of spine of scapula	3	5
Vertical diameter of proximal articular surface of fore-arm	3	6
Transverse diameter of proximal articular surface of fore-arm	3	5
Height of olecranon	5	3

[26] This diameter increases rapidly in the posterior lumbar vertebrae, in correspondence with the enlargement of the spinal cord, which gives off the great nerves of the hinder extremities.

ADMEASUREMENTS OF THE BONES OF THE MACRAUCHENIA – *continued*

	Inches	*Lines*
Greatest diameter of its base	2	–
Circumference of proximal end of anchylosed radius and ulna	11	10
Entire length of inner toe of fore-foot, inclusive of metacarpal bone	13	–
Breadth of proximal end of metacarpus	3	8
Breadth of distal end of metacarpus	5	4
Length of inner metacarpal bone	7	6
Length of middle metacarpal bone	8	–
Length of outer metacarpal bone	7	–
Length of inner proximal phalanx	3	6
Length of middle proximal phalanx	2	10
Length of outer proximal phalanx	3	4
Length of inner middle phalanx	2	–
Length of middle phalanx	2	3
Length of inner distal phalanx[27]	1	–
Length of the femur	24	–
Diameter of base of articular surface of the head of the femur	3	6
Greatest diameter of proximal end	7	–
Greatest diameter of distal end	6	3
Circumference of middle of shaft	8	–
Length of tibia	18	–
Greatest diameter of proximal end	5	7
Greatest diameter of distal end, including fibula	4	4
Circumference of middle of shaft	9	–
Length of metatarsal bone[28]	7	4 /

DESCRIPTION OF A FRAGMENT OF A CRANIUM OF AN EXTINCT MAMMAL INDICATIVE OF A NEW GENUS OF EDENTATA, AND FOR WHICH IS PROPOSED THE NAME OF

GLOSSOTHERIUM

'La première chose à faire dans l'étude d'un animal fossile, est de reconnaitre la forme de ses dents molaires; on détermine par-là s'il est carnivore ou herbivore;' says Cuvier, at the commencement of that series of splendid chapters in which the restoration of the extinct Pachyderms of the Paris Basin is recorded. In the present case,

[27] The relative breadth of these bones is shown in the figures of the fore-foot, Pl. XI.
[28] The figures in Pl. XIV preclude the necessity of giving the admeasurements of the astragalus.

however, as in that of the Mammiferous animal whose fossil remains we were last considering, the important organs, to which Cuvier directs our first attention, are wanting. Nor are there here, as in the Macrauchenia, any remains of the locomotive extremities to compensate for the deficiency of teeth, and guide us into the right track of investigation and comparison. The animal, the nature and affinities of which are the subject of the following pages, is, in fact, represented in Mr Darwin's collection, by nothing more than a fragment of the cranium.

This fragment, which was found in the bed of the same river (see p. [16] 3) in Banda Oriental, with the cranium of the Toxodon, includes the parietes of the left side of the cerebral cavity, the corresponding nervous and vascular foramina, the left occipital condyle, a portion of the left zygomatic process, and, fortunately also, the left articular surface for the lower jaw. The importance of this surface in the determination of the affinities of a fossil animal has been duly appreciated, since the relations of the motions of the lower jaw to the kind of life of each animal were pointed out by Cuvier; but yet we should be deceived were we to establish, in conformity with the generalization enunciated by Cuvier,[29] our conclusion, from this surface, of the nature of the food of the extinct species under consideration; / for the glenoid cavity is so shaped as to allow the lower jaw free motion in a horizontal plane, from right to left, and forwards or backwards, like the movements of a mill-stone; and, nevertheless, I venture to affirm it to be most probable, that the food of *Glossotherium* was derived from the animal and not from the vegetable kingdom; and to predict, that when the bones of the extremities shall be discovered, they will prove the Glossothere to be not an ungulate but an unguiculate quadruped, with a fore-foot endowed with the movements of pronation and supination, and armed with claws, adapted to

[29] 'Comme le genre de vie de chaque animal est toujours en rapport avec les mouvements dont sa mâchoire est susceptible, on retrouve dans la conformation des surfaces destinées à l'articulation, les particularités qui semblent le déterminer d'avance. Ainsi dans les animaux qui vivent de chairs, substances filamenteuses qui ne peuvent être écrasées, mais seulement coupées et dechirées, le mouvement de la mâchoire inférieure ne peut s'exécuter que de haut én bas. Dans les herbivores, les frugivores et les granivores, comme le principal mouvement est celui de broiement pour écraser, comprimer les herbes et les fruits, pour briser les grains et les réduire, en pâte, le movement des mâchoires se fait encore de droite à gauche, et réciproquement, on en même temps, de devant en arrière, en un mot, dans un plan horizontal autant que dans un vertical: les uns représentent des ciseaux, les autres des meules de moulin.'

make a breach in the strong walls of the habitations of those insect-societies, upon which there is good evidence in other parts of the present cranial fragment, that the animal, though as large as an ox, was adapted to prey.

We perceive, in the first place, looking upon the base of this portion of skull, a remarkable cavity, situated immediately behind the tympanic bone, of nearly a regular hemispherical form, an inch in diameter (fig. 2, *b*, Pl. XVI). The superficies of this cavity appears not to have been covered with articular cartilage, for it is irregularly pitted with many deep impressions; and I conclude, therefore, that it served to afford a ligamentous attachment to the styloid element of a large *os hyoides*. With this indication of the size of the skeleton of the tongue, is combined a more certain proof of the extent of its soft, and especially its muscular parts, in the magnitude of the foramen, for the passage of the lingual or motor nerve (*c*, fig. 2 and 3). This foramen (the anterior condyloid), in the present specimen, is the largest of those which perforate the walls of the cranium, with the exception of the foramen magnum; it is fully twice the size of that which gives passage to the second division of the fifth nerve; its area is oval, and eight lines in the long diameter, so that it readily admits the passage of the little finger.

It is only in the Ant-eaters and Pangolins that we find an approximation to these proportions of the foramen for the passage of the muscular nerve of the tongue; and the existing Myrmecophagous species even fall short of the larger fossil in this respect. Some idea of the size of the lingual nerve, and of the organ it was destined to put in motion, may be formed, when it is stated that the foramen giving passage to the corresponding nerve in the Giraffe – the largest of the Ruminants, and having the longest and most muscular tongue in that order – is scarcely more than one-fourth the size.

With these indications of the extraordinary development of the tongue, we are naturally led, in order to carry out a closer and more detailed comparison of the fossil in question, to that group of mammalia in which the tongue plays the chief part in the acquisition of the food. The size, form, and position of the occipital condyle – the magnitude of the occipital foramen (which must here have somewhat exceeded three inches in the transverse diameter) – the slope of the occipital surface of the cranium from below, upwards and forwards, at an angle of 60° / with the base of the cranial cavity – each and all attest the close affinities of the present animal to the Edentata. More decisive evidence of the same relationship will be adduced from the organization of other parts of the cranium. The glenoid articular surface (*a*, fig. 2, Pl. XVI) is an almost flattened plane, wider in the transverse than in the longitudinal direction; and, as in the genera *Myrmecophaga* and *Manis*, it is not defended behind by any descending process. In its general form it resembles the glenoid cavity of *Orycteropus* more than that of the preceding Edentates; but, in *Orycteropus*, the articulation is defended posteriorly by a descending process of the zygoma, and it is also situated relatively closer to the os tympanicum.

Had the *Glossotherium* teeth? The extent of the temporal muscle, which is indicated by the rugged surface of the temporal fossa, and by the well-marked boundary, formed by a slightly elevated bony ridge, which extends to near the line of the sagittal suture, together with the size of the zygomatic portion of the temporal bone, and the remains of the oblique suture by which it was articulated to the malar bone, enables me to answer this question confidently in the affirmative. They will probably be found to be molar teeth of a simple structure, as in the Orycteropus.

The evidence just alluded to of the existence of an os malae is interesting, because this bone is wanting in the Pangolins; and its rudimental representative in the true Ant-eaters does not reach the zygomatic process of the temporal bone, which consequently has no articular or sutural surface at its anterior extremity. In the presence, therefore, of the surface for the junction of the os malae, and the consequent evidence of the completion of the zygomatic arch, we learn that the Glossothere was more nearly allied to the Armadillos and Orycterope. That its affinity to the latter genus was closer than to the Armadillos we have most interesting evidence in the form and loose condition of the tympanic bone: it is represented of the natural size at fig. 4, Pl. XVI. Through the care and attention devoted to his specimens by their gifted discoverer, this bone was preserved *in situ*, as represented at *d*, fig. 1; but it had no osseous connection with the petrous or other elements of the temporal bone, and could be displaced and replaced with the same ease as in the *Orycterope*. This bony frame of the membrana tympani, in the Glossothere, describes rather more than a semicircle, having the horns directed upwards; it has a groove, one line in breadth, along its concave margin, for the attachment of the ear-drum, and sends down a rugged process, half an inch long, from its lower margin. In the *Dasypodes* and *Myrmecophagae*, the tympanic bone soon becomes anchylosed with the other parts of the temporal; it is only in *Orycteropus*, among the existing insectivorous

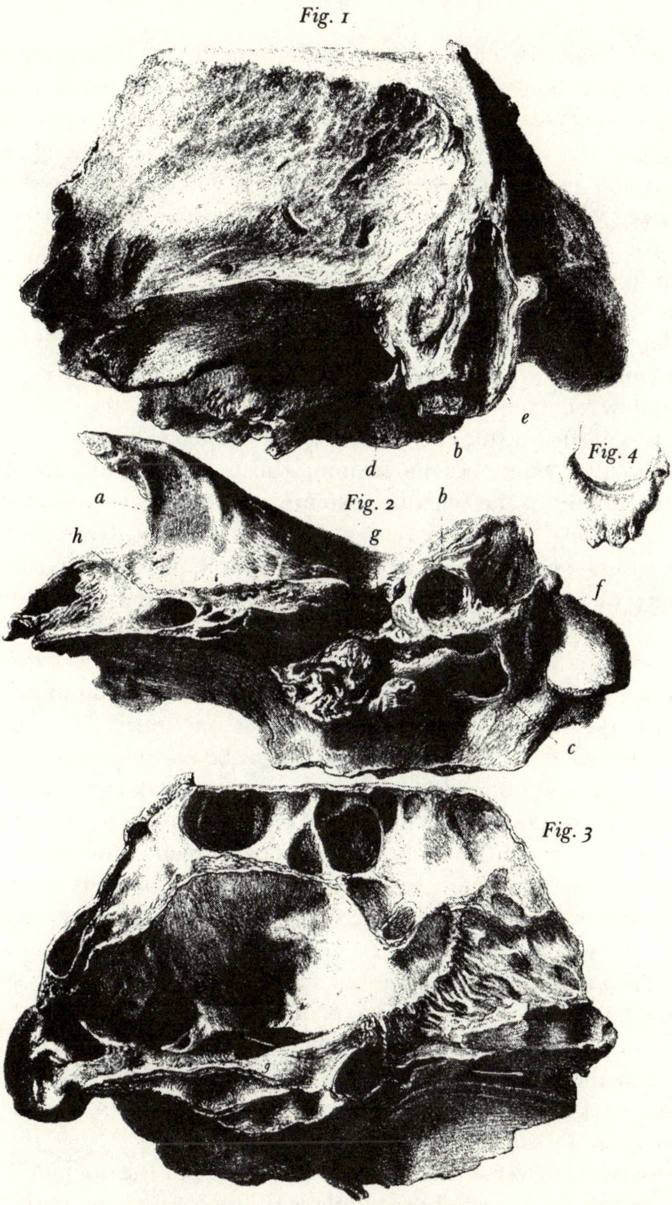

PLATE XVI

Fragment of the cranium of the Glossotherium

Bruta or *Edentata*, that it manifests throughout life the foetal condition of a distinct bony hoop, deficient at the upper part. The os tympanicum of *Orycteropus*, however, differs from that of / *Glossotherium*, in forming part of the circumference of an ellipse, whose long axis is vertical; and in sending outwards, from its anterior part, a convex eminence, which terminates in a point directed downwards and forwards.

Such appear to be the most characteristic features of the cranial fragment under consideration, in which we have found, that the articular surface for the os hyoides throws more light upon the nature of the animal of which it is a part, than even the glenoid cavity itself. There now remains to be described as much of the individual characters of the constituent bones as the specimen exhibits.

The occipital bone, besides forming the posterior and part of the inferior parietes of the cranium, extends for about half an inch upon the sides, where the ex-occipital element is articulated by a vertical suture with the mastoid element of the temporal: this suture is situated in a deep and well-marked muscular depression (*e*, fig. 1), measuring three inches in the vertical, and upwards of one inch in the transverse direction. The other sutures, uniting the occipital to the adjoining bones, are obliterated. The breadth of the occipital region must have exceeded the height of the same by about one-third. The condyle extends nearly to the external boundary of the occipital aspect of the cranium; there is situated, external to it, only a small ovate, rounded and smooth protuberance. The slightly concave surface of the occipital plane of the cranium is bounded above by a thick obtuse ridge, the muscular impressions are well sculptured upon it. It is traversed transversely at its upper third by a slighly elevated bony crest; and the surface below this ridge is again divided by a narrower intermuscular crest, which runs nearly vertically, at about an inch and a half from the external boundary of the occipital plane. As a similar crest must have existed on the opposite side, the general character of the occipital surface in the Glossothere would resemble that of the Toxodon. A similar correspondence may be noticed in the terminal position of the condyle, and the slope of the occipital plane.

Above the transverse ridge, the rough surface of the occipital plane slopes forward, at a less obtuse angle with the basal plane, to the first named ridge which separates the occipital from the coronal or superior surface of the skull. The contour of this surface runs forwards, as far as the fragment extends, in an almost straight line: the extent of surface between the temporal muscular ridges must have been about five inches posteriorly, but it decreases gradually as it extends forwards: all that part which is preserved is quite smooth. The attachment of the fasciculi of the temporal muscle, and the convergence of their fibres as they passed through the zygoma are well marked on the sculptured surface of the bone. The zygomatic process is relatively stouter than in *Orycteropus*: it is prismatic: the external facet is nearly plane: the superior is concave, and increases in breadth anteriorly: the inferior surface offers a slight convexity behind the flattened articular surface for the lower jaw. The margin / of the zygoma formed by the meeting of the upper and lower facets presents a semicircular curve, extended transversely from the cranium, and directed forwards.

The anterior extremity is obliquely truncated from below upwards and forwards, and presents a flattened triangular surface indicative of its junction with an os malae: the space between this extremity and the side of the cranium measures one inch and nine lines across, and thus gives us the thickness of the temporal muscle. The distance from the origin of the zygoma to the occipital plane is relatively greater than in *Orycteropus*; *Glossotherium* is in this respect more similar to *Myrmecophaga* and *Manis*.

The sphenoid bone forms a somewhat smooth protuberance below and behind the base of the *zygoma*. The tympanic bone is wedged in between this protuberance in front, and the mastoid process behind. The chief peculiarity of the broad mastoid is the regular semicircular cavity at its under part for the articulation of the styloid bone of the tongue. This depression is separated below by a broad rough protuberance from the foramen jugulare (*f*, fig. 2, Pl. XVI), which is immediately external to, and slightly in advance of the great foramen condyloideum, *c*. A small rugged portion of the os petrosum separates the jugular from the carotid canal, which arches upwards and directly inwards to the side of the shallow sella turcica (the external and internal orifices of the carotid canal are shown at *g*, figs 2 and 3). The chief protuberance on the basis cranii is a large and rugged one, serving for the attachment of muscles, and due chiefly to the expansion of a great sinus in the body of the sphenoid. This protuberance is separated from the smaller sphenoid protuberance before mentioned by a large groove continued downwards and forwards from the tympanic cavity, and containing the Eustachian tube, which does not traverse a complete osseous canal. Immediately internal to the glenoid cavity is the large orifice of the canal transmitting the third division of the fifth pair of nerves, the principal branch of which endows the tongue with sensibility; this foramen (*h*, fig. 2) is rather less than that for the muscular nerve of the tongue.

The internal surface of the present cranial fragment affords a very satisfactory idea of the size and shape of the brain of the extinct species to which it belongs. It is evident that, as in other Bruta, the cerebellum must have been almost entirely exposed behind the cerebrum; and that the latter was of small relative size, not exceeding that of the Ass; and chiefly remarkable, as in the Orycterope, Ant-eater, and Armadillo for the great development of the olfactory ganglia. The antero-posterior extent of the cribriform plate, as exposed in this fragment, is three inches, and the complication of the oethmoid olfactory lamellae which radiate from it into the nasal cavity is equal to that which exists in the smaller Edentata (fig. 3, Pl. XVI). The nasal cavity is complicated in *Glossotherium* / by the great number and capacious size of the air-cells which are in communication with it: these extend over all the upper, lateral, and back parts of the cranial cavity, as far even as the upper boundary of the foramen magnum: they also occupy the anterior two-thirds of the basis cranii. The external configuration of the skull would, therefore, afford a very inadequate or rather deceptive notion of the capacity of the cerebral cavity, were not the existence and magnitude of these sinuses known. The interspace of the outer and inner tables of the cranium are separated above the origins of the olfactory ganglia for the extent of three inches: above the middle of the cerebrum they are an inch and a half apart; at the sides of the cranium the interposed air-cells are from one to two inches across; at the back part of the cranium about one inch. The sinuses have generally a rounded form.

The foramen rotundum (through which in figure 3 a probe is represented as passing), and the foramen ovale are situated close together, within a common transversely oblong depression (*i*). The carotid canal (*g*) opens into the outer side of the commencement of this wide channel, which conducts the great fifth pair of nerves to the outlets of its two chief divisions.

The petrous bone projects into the cranial cavity, in the form of an angular process with three facets: the foramen auditorium internum (*k*), and the aqueductus vestibuli, are situated on the posterior facet. Immediately behind the os petrosum is the foramen lacerum jugulare (*l*), situated at the point of convergence of the vertical groove of the lateral sinus, with a groove of similar size continued forwards from above the anterior condyloid canal. The plane of the internal opening of this canal (*c*, fig. 3) is directed obliquely inwards and backwards, and the lateral wall of the foramen magnum behind the foramen condyloideum slopes outwards to the edge of the condyle. Immediately internal to the foramen condyloideum is a small vascular foramen conducting a branch of the basilar artery into the condyloid canal, for the nourishment, doubtless, of the great lingual nerve.

In the relations of the plane of the internal orifice of the anterior condyloid foramen with that of the foramen magnum, we search in vain for a corresponding structure in any of the Mammiferous orders, save the Edentata;[30] and among these the Orycterope comes nearest the Glossothere in this respect. In the degree of development of the internal osseous ridge giving attachment to the tentorium cerebelli, the Ant-eaters and Armadillos more resemble the Glossothere than does the Orycterope; in which a continuous bony plate arches across the cranial cavity: in the Manis a still greater proportion of the tentorium is ossified, / and it consequently recedes the furthest amongst the Edentata, in this, as in most other particulars of the cranial organization, from the Glossothere. The chief distinctive peculiarity in the cranium of the Glossothere, so far as it can be studied in the present fragment, and compared with that of other Edentata, is the deep, well-marked, semicircular styloid depression, above described.

A question may arise after perusing the preceding evidence, upon which the present fossil is referred to a great Edentate species nearly allied to the *Orycteropus*, whether one or other of the lower jaws, subsequently to be described, and in like manner referable, from their dentition, either to the *Orycteropodoid* or *Dasypodoid* families of Edentata, may not have belonged to the same species as does the present mutilated cranium. I can only answer, that those jaws were discovered by Mr Darwin in a different and very remote locality – that no fragments or teeth referable to them were found associated with the present fossil; and that, as it would be, therefore, impossible to determine from the evidence we have now before us, which of the two lower jaws should be associated with *Glossotherium*; and as both may with equal if not greater probability belong to a totally distinct genus, it appears to me to be preferable, both in regard to the advancement of our knowledge of these most interesting Edentata of an ancient world,

[30] In the monotrematous Echidna, the large canal for the lingual nerve has a widely different direction and course from that in the placental Edentata.

as well as for the convenience of their description, to assign to them, for the present, distinct generic appellations.

The figures in Plate XVI preclude the necessity of a table of admeasurements of the cranial fragment of *Glossotherium*.

DESCRIPTION OF A MUTILATED LOWER JAW AND TEETH, ON WHICH IS FOUNDED A SUBGENUS OF MEGATHERIOID EDENTATA, UNDER THE NAME OF

MYLODON

The genus *Megalonyx*, as is well known, owes its name and the discovery of the fossil remains on which it was founded, to the celebrated Jefferson,[31] formerly President of the United States. Cuvier, from an examination of a single tooth, and the casts of certain bones of the extremities, especially the terminal ones, determined the ordinal affinities of this remarkable extinct quadruped.[32] But while he / retained the name of *Megalonyx*, and used it in a generic case, Cuvier offered no characters whereby other fossil remains might be generically either distinguished from, or identified with the *Megalonyx Jeffersonii*, unless, among such remains there happened to be a tooth, or a claw exactly corresponding with the descriptions and figures in the *Ossemens Fossiles*; and when, of course, a specific identity, and not merely a generic relationship would be established.

The greater part of Cuvier's chapter on *Megalonyx* is devoted to the beautiful and justly celebrated reasoning on the ungueal phalanx, whereby it is proved to belong, not to a gigantic Carnivore of the Lion-kind, as Jefferson supposed, but to the less formidable order of Edentate quadrupeds; and Cuvier, in reference to the tooth – the part on which alone a generic character could have been founded – merely observes that it resembles at least as much the teeth of one of the great Armadillos, as it does those of the Sloths.[33]

[31] *Transactions of the Philosophical Society of Philadelphia*, vol. iv, p. 246.

[32] Its relations to the Edentata, previously conjectured by Dr Wistar, are proved in the *Annales du Muséum*, tom. v, p. 358; its more immediate affinities as an annectant form in that group are discussed in the edition of the *Ossem. Fossiles*, of 1833, tom. v, pt 1, p. 160.

[33] Speaking of this tooth, Cuvier observes, 'Je l'avois cru d'abord nécessairement *de paresseux*; mais aujourd-hui que je connois mieux l'ostéologie des divers tatous, je

In the last edition of the *Régne Animal,* Cuvier introduces the *Megatherium* and *Megalonyx,* between the Sloths and Armadillos; but alludes to no other difference between the two genera than that of size – 'l'autre, le *Megalonyx,* est un peu moindre' (p. 226). Some systematic naturalists, as Desmarest, and Fischer, have, therefore, suppressed the genus, and made the *Megalonyx* a species of *Megatherium* under the name of *Megatherium Jeffersonii.* The dental characters of the genus *Megatherium* are laid down by Fischer[34] as follows: *Dent. prim. et lan.* % *molares* 4/4–4/4, *obducti, tritores, coronide nunc planâ transversim sulcatâ nunc medio excavatâ marginibus prominulis.*' That *Megalonyx* had the same number of molares as *Megatherium* (supposing that number in the Megathere to be correctly stated, which it is not), is here assumed from analogy, for neither Jefferson, Wistar, nor Cuvier – the authorities for *Megalonyx* quoted by Fischer – possessed other means of knowing the dentition of that animal than were afforded by the fragment of a single tooth.

Now the almost entire lower jaw about to be described offers, in so far as respects the general form and structure of the teeth, the same kind and degree of correspondence with the *Megatherium,* as does the *Megalonyx Jeffersonii* of Cuvier; and, what is only probable in that species, is here certain, viz., an agreement with the Megatherium in the class, viz. *molares,* to which the teeth exclusively belong. The question, therefore, on which I find myself, in the outset, called upon to come to a decision is, as to the preference of the mode of viewing the subject of the generic relationship of the *Megalonyx* adopted by Desmarest, Fischer, etc., or of that, on which Cuvier, and after him Dr Harlan, have practically acted: whether, in short, the genus *Megatherium* is to rest upon the more / comprehensive characters of kind and general structure of the teeth, or upon the more restricted ones, of form and such modifications in the disposition and proportions of the component textures of the tooth, as give rise to the characteristic appearances of the triturating surface of the crown.

With the respect to existing Mammalia, most naturalists of the present day seem to be unanimous as to the convenience at least of founding a generic or sub-generic distinction on well marked modifications in the form and structure of the teeth, although they

[34] *Synopsis Mammalium.*

trouve qu'elle ressemble au moins autant à une dent de l'un des grands *tatous.* – Loc. cit., p. 172.

may correspond in number and kind, in proof of which it needs only to peruse the pages of a *Systema Mammalium* which relate to the distribution of the Rodent Order. According to this mode of viewing the logical abstractions under which species are grouped together, the extinct Edentate Mammal discovered by Jefferson must be referred to a genus distinct from *Megatherium*, and for which the term *Megalonyx* should be retained. This will be sufficiently evident by comparing the descriptions given by Cuvier of one of the teeth of the *Megalonyx Jeffersonii*, and by Dr Harlan of a tooth of his *Megalonyx laqueatus*, with those of the *Megatherium* which have been published by Mr Clift. The fragment of the molar tooth of the *Megalonyx Jeffersonii*, described and figured in the *Ossemens Fossiles*, seems to have been implanted in the jaw, like the teeth of the *Megatherium*, by a simple hollow base similar in form and size to the protruded crown: its structure Cuvier describes as consisting of a central cylinder of bone enveloped in a sheath of enamel.[35] The transverse section of this tooth presents an irregular elliptical form, the external contour being gently and uniformly convex, the internal one, undulating; convex in the middle, and slightly concave on each side, arising from the tooth being traversed longitudinally on its inner side by two wide and shallow depressions.

The imperfect tooth of the species called by Dr Harlan *Megalonyx laqueatus*, and of which a cast was presented by that able and industrious naturalist to the Museum of the Royal College of Surgeons, resembles in general form, and especially in the characteristic double longitudinal groove on the inner side, the tooth of the *Megalonyx Jeffersonii*. It is thus described by Dr Harlan:

'The fractured molar tooth appears to have belonged to the inferior maxilla on the right side; the crown is destroyed; a part of the cavity of the root remains. The body is compressed transversely, and presents a double curvature, which renders its anterior and exterior aspects slightly convex; the posterior and interior gently concave; these surfaces are all uniform, with the exception of the interior or mesial aspect, which presents a longitudinal rib or ridge, one-half the thickness of the long diameter of the tooth; with a broad, not profound longitudinal / groove or channel along each of its borders. It

[35] It is most probable that the substance which is here termed 'enamel', is similar to that which forms the dense prominent ridges in the tooth of the Megatherium, and which I have shown to be composed of minute parallel calcigerous tubes, similar to the ivory bone of the human tooth.

is from this resemblance to a portion of a fluted column, that the animal takes its specific appellation (*Meg*^{x}. *laqueatus*).

'The crown would resemble an irregular ellipsis widest at the anterior portion. The tooth consists of a central pillar of bone surrounded with enamel, the former of a dead white, the latter of a ferruginous brown colour: the transverse diameter is more than two-thirds less than its length, whilst that of *Meg*^{x}. *Jeffersonii* is only one-third less – the antero-posterior diameter is one-half its length in the former, and two-thirds less in the latter. The proportions of this tooth are consequently totally at variance with that of its kindred species.' [Vide Pl. XII, fig. 7, 8, 9.][36]

Dr Harlan describes also two claws of the fore-foot, a radius, humerus, scapula, one rib, an os calcis, a metacarpal bone, certain vertebrae, a femur, and tibia, of the same *Megalonyx*; these parts of the skeleton, together with the tooth, which so fortunately served to establish the generic relationship of the species with the *Megalonyx* of Jefferson and Cuvier, were discovered in Big-bone-cave, Tennessee, United States.

Dr Harlan does not enter into the question of the generic characters of *Megalonyx*, but it would seem that he felt them to rest not entirely on dental modifications, for he observes that 'a minute examination of the tooth and knee-joint renders it not improbable, supposing the last named character to be peculiar to it, that if the whole frame should hereafter be discovered, it may even claim a generic distinction, in which case, either Aulaxodon, or PLEURODON, would not be an inappropriate name.'[37]

There can be no doubt, as it appears to me, with respect to a fossil jaw presenting teeth in the same number, and of the same general structure, as in the *Megatherium*, and with individual modifications of form, as well marked as those which distinguish *Megatherium* from *Megalonyx*, that the Palaeontologist has no other choice than to refer it, either as Fischer has done with *Megalonyx*, to a distinct species of the genus *Megatherium*, or to regard it as the type of a subgenus distinct from both. With reference, however, to the *Pleurodon* of Dr Harlan, after a detailed comparison of the cast of the tooth on which that genus is mainly founded, with the descriptions and figures of the tooth of the *Megalonyx Jeffersonii*, in the *Ossemens Fossiles*, they seem to differ in so

[36] *Medical and Physical Researches*, pp. 323–4.
[37] Loc. cit., p. 330.

slight a degree as to warrant only a specific distinction, and this difference even, viewing the various proportions of the teeth in the same jaw of the *Megatherium*, is more satisfactorily established by the characters pointed out by Dr Harlan in the form and proportion of the radius, than by those in the tooth itself. /

The next notice of the *Megalonyx* which I have consulted, in the hope of meeting with additional and more precise information as to its real generic characters, is an account given by the learned Professor Doellinger,[38] of some fossil bones, collected by the accomplished travellers Spix and Martius in the cave of Lassa Grande, near the Arrayal de Torracigos, in Brazil. In this collection, however, it unfortunately happens that there are no teeth, but only a few bones of the extremities, including some ungueal phalanges, which Professor Doellinger concludes, from their shape, the presence of an osseous sheath for the claw, and the form of their articulation, to belong, without doubt, to an animal of the Megatherioid kind, about the size of an Ox. He particularly states that they are not bones of an immature individual; but that they agree sufficiently with Cuvier's descriptions and figures of the *Megalonyx* to be referred to that species of animal (zu dieses thierart); and he adds, what is certainly an interesting fact, that the fossils in question form the first of the kind that had been discovered out of North America.

Subsequently to the discovery of these bones, and of those of the *Megalonyx laqueatus* above alluded to, the remains of another great Edentate animal were found in North America, and were deposited in the Lyceum at New York; among these is a portion of the lower jaw with the whole dental series of one side. It is thus described by Dr Harlan.

'The fragment I am now about to describe is a portion of the dexter lower jaw of the Megalonyx, containing four molar teeth; three of the crowns of these teeth are perfect, that of the anterior one is imperfect. These teeth differ considerably from each other in shape, and increase in size from the front, the fourth and posterior tooth being double the size of the first, and more compressed laterally; it is also vertically concave on its external aspect, and vertically convex on its internal aspect; the interior or mesial surface is strongly fluted, and it has a deep longitudinal furrow on the dermal aspect, in which respect it differs from the tooth of the *M. laqueatus* previously described by me,

[38] Spix and Martius, *Reise in Brazil*, Band ii, p. 5.

of which the dermal aspect is uniform, but to which, in all other respects, it has a close resemblance. I suppose it therefore probable, that this last may have belonged to the upper jaw. The three anterior molars differ in shape and markings: they are vertically grooved, or fluted, on their interior and posterior aspects, a transverse section presenting an irregular cube. The length of the crown of the posterior molar is two inches: the breadth about five-tenths of an inch: the length of the tooth is three inches and six-tenths. The diameter of the penultimate molar is eight-tenths by seven-tenths of an inch. The length of this fragment of the jaw-bone is eight inches and four-tenths; the height three inches and six-tenths: the length of the space occupied by the alveolar sockets five inches and eight-tenths. / The crown of the tooth presents no protuberances, but resembles that of the Sloth; the roots are hollow.'[39]

This fossil is referred by Dr Harlan to his *Megalonyx laqueatus*; but, pending the absence of other proof of the identity of species, in which, as may be seen by comparing fig. 2, with fig. 4, in Pl. XVII, the teeth differ widely in form, it would be obviously hazardous to adopt such an approximation on hypothetical grounds.[40] In order, however, to obtain more satisfactory evidence of the nature and amount of the difference between the *Megalonyx laqueatus*, and the allied animal represented by the above-described fragment of lower jaw, I wrote to my much respected friend M. Laurillard, requesting him to send me a sketch of the teeth in the cast of that lower jaw, which had been transmitted from New York to the Garden of Plants. With full confidence in the characteristic precision and accuracy of the drawing

[39] Harlan, *Medical and Physical Researches*, 1835, p. 334. M. de Blainville speaks of a cast of a fragment of a lower jaw 'portant encore *cinq* dents en série'; as having been transmitted to the Museum of the Garden of Plants from North America, together with other bones, all of which he refers to the genus *Megalonyx*; M. de Blainville does not describe these teeth, which is to be regretted, inasmuch as, if he be correct in regard to their number, which can hardly be doubted, and if he wrote with any clear and definite ideas of the generic characters of *Megalonyx*, this would indicate that *Megalonyx* differed generically both from *Megatherium* and *Mylodon* in a more important dental character than has hitherto been suspected (See *Comptes Rendus*, etc., 1839, No. V, p. 142).

[40] Dr Harlan also indicates differences in certain parts of the skeleton of the New York fossils as compared with his *Megx. laqueatus*; but thinks them probably due to a difference in the age of the individuals: he says 'There is also in Mr Graves' collection, in New York, a tibia, nearly perfect from the right leg; the segment of a flattened sphere, on which the external condyle of the femur moves, is rather more depressed, than in the specimen from Big-bone-cave. Other marks and peculiarities are observable on this bone, not found on that of the *Megalonyx laqueatus* of Big-bone-cave, but they are probably due to a difference in the age of the individuals.' Loc. cit., p. 335.

with which I have been obligingly favoured by M. Laurillard, I am disposed to regard the amount of difference recognizable in every tooth in the lower jaw in question (figs 3 and 4) as compared with the molar tooth either of *Megalonyx Jeffersonii* (fig. 1), or *Meg*^{x}. *laqueatus* (fig. 2) to be such as to justify its generic separation from *Megalonyx* on the same grounds as *Megalonyx* is distinguished from *Megatherium*, and for the subgenus of Megatherioid Edentata, thus indicated, I would propose the name of MYLODON.[41] The species of which the fossil remains are described by Dr Harlan may be dedicated to that indefatigable Naturalist who has contributed to natural science so much valuable information respecting the Zoology, both recent and fossil, of the North American continent. The fossil about to be described represents a second and smaller species of the same genus, and I propose to call it *Mylodon Darwinii*, in honour of its discoverer, of whose researches in the Southern division of the New World it forms one of many new and interesting fruits. /

This fossil was discovered in a bed of partly consolidated gravel at the base of the cliff called Punta Alta, at Bahia Blanca in Northern Patagonia: it consists of the lower jaw with the series of teeth entire on both sides: but the extremity of the symphysis, the coronoid and condyloid processes, and the angular process of the left ramus, are wanting. The teeth are composed, as in *Bradypus*, *Megatherium* and *Megalonyx*, of a central pillar of coarse ivory, immediately invested with a thin layer of fine and dense ivory, and the whole surrounded by a thick coating of cement.

In the fig. 5, Pl. XVII, the fine ivory is represented by the white striated concentric tract on the grinding surface of the teeth; it is of a yellowish-white colour in the fossil, and stands out, as an obtuse ridge, from that surface: both these conditions depend on the large proportion of the mineral to the animal constituent in this substance of the tooth. The external layer of the cement presents in the fossil the same yellowish-brown tint as the bone itself, which it so closely resembles, both in intimate structure and in chemical composition; the internal layer next the dense ivory is jet black, indicating the great proportion of animal matter originally present in this part. The central pillar of coarse ivory, which, from its more yielding texture, has been worn down into a hollow at the triturating surface of the tooth, also presents, as a consequence of the less proportion of the hardening

[41] Μυλη, *mola*; οδους, *dens*.

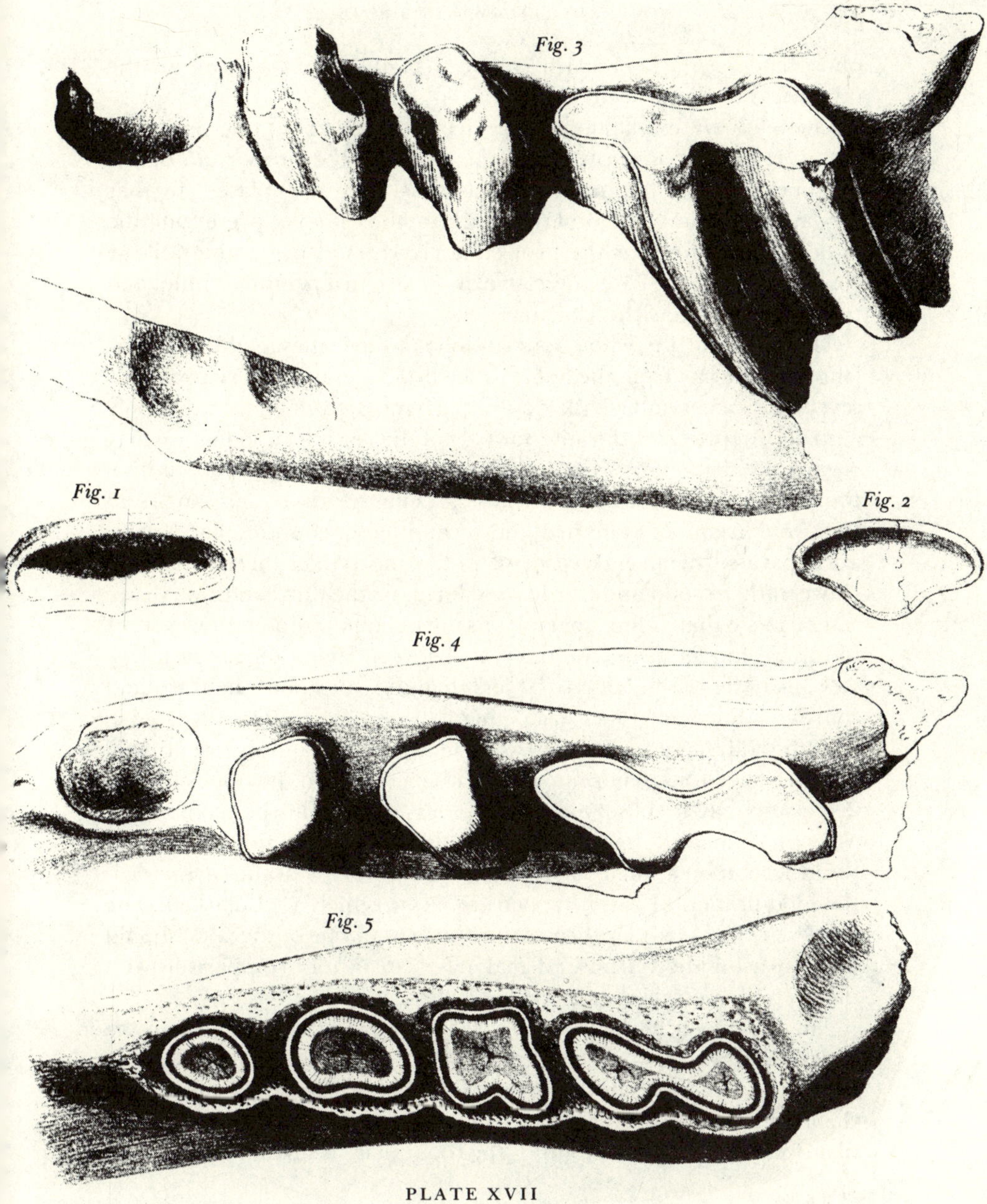

PLATE XVII

Fig. 1. Megalonyx Jeffersonii *Fig. 2. Megalonyx laqueatus*
Figs 3, 4. Mylodon Harlani *Fig. 5. Mylodon Darwinii*

phosphates, a darker brown colour than the external layer of the cement, or the bone itself.

The teeth are implanted in very deep sockets; about one-sixth only of the last molar projects above the alveolus; the proportion of the exposed part of the tooth increases as they are placed further forwards. The implanted part of each tooth is simple; preserving the same size and form as the projecting crown, and presenting a large conical cavity at the base, indicative of the original persistent pulp, and perpetual growth of these teeth.

The extent of the whole four alveoli is four inches, eight lines; the length of the jaw from the angle to the broken end of the symphysis is seventeen inches and a half;[42] from the figures it will be seen that only a small proportion of the anterior part of the jaw is lost, so that we may regard the dentigerous part of the jaw as being limited to about one-fourth of its entire length; the alveoli being nearly equidistant from the two extremities. The first and second teeth, counting backwards, are separated by an interspace of rather more than three lines; that between the second and third is one line less; the third and fourth are rather more than a line apart: from the oblique position, however, of the three hinder teeth the intervals between them appear in a side view, as in fig. 1, Pl. XIX, to be less than in reality, and the third and fourth teeth seem to touch each other. /

Each tooth has a form and size peculiar to itself, and different from the rest, but corresponds of course with its fellow on the opposite side. The same may be observed, but in a less degree, in the teeth of the Megatherium itself; hence, it is obviously hazardous to found a generic distinction upon a single tooth, unless, as in the case of the *Glyptodon*,[43] the modification of form happens to be extremely well marked. The whole series of teeth, or their sockets, at least of one of the jaws, should be known for the purpose of making a satisfactory comparison with the previously established Edentate genera.

The first molar in the present jaw is the smallest and simplest of the series: its transverse section is ellipsoid, or subovate, narrowest in front, and somewhat more convex on the outer than on the inner side: the long diameter of the ellipse is nine lines, the short or transverse diameter six lines: the length of the tooth may be about three inches,

[42] If the lower jaw of *Mylodon Harlani*, bears the same proportion to its teeth as does that of *Mylodon Darwinii*, it must be about two feet in length.

[43] See *Proceedings of the Geological Society*, March 1839, and Parish, *Buenos Ayres*, p. 178, *b*, Pl. 1, fig. 2 and 3.

but I have not deemed it necessary to fracture the alveolus in order to ascertain precisely this point.

The second tooth presents in transverse section a more irregular and wider oval figure than the first: the line of the outer side is convex, but that of the inner side slightly concave, in consequence of the tooth being traversed longitudinally by a broad and shallow channel or impression; the longitudinal diameter of the transverse section is one inch; the transverse diameter at the widest part nine lines. There is a slight difference in the size of this tooth on the two sides of the jaw, the right one, from which the above dimensions are taken, being the largest.

The transverse section of the third tooth has a trapezoidal or rhomboidal form; the angles are rounded off; the posterior one is most produced; the anterior and posterior surfaces are flattened, the latter slightly concave in the middle; the external and internal sides are concave in the middle, especially the inner side, where the concavity approaches to the form of an entering notch. The longest diameter of the transverse section of this tooth is thirteen lines, the shortest seven lines and a half: in the tooth on the right side the external surface is nearly flat; this slight difference is not indicated in the figure (Pl. XVIII).

The last molar, which is generally the most characteristic in the fossil *Bruta*, presents in an exaggerated degree the peculiarities of the preceding tooth; the longitudinal channels on both the outer and inner surfaces encroach so far upon the substance of the tooth, that the central coarse ivory substance is as it were squeezed out of the interspace, and the elevated ridge of the dense ivory describes an hour-glass figure upon the triturating surface, the connecting isthmus being but half the breadth of the rest of the tract; the external caementum preserves nearly an equal thickness throughout. Of the two lobes into which this tooth is / divided by the transverse constriction, the anterior is the largest; their proportions and oblique position are pretty accurately given in the figure. The longitudinal diameter of the transverse section of this tooth is one inch, seven lines, its greatest lateral or transverse diameter is ten lines, its least diameter at the constricted part is three lines, the length of the entire tooth is four inches. Judging from the form of the jaw, the length of the other teeth decreases in a regular ratio to the anterior one. The posterior tooth is slightly curved, as shown in fig. 2, Pl. XIX, with the concavity directed towards the outer side of the jaw.

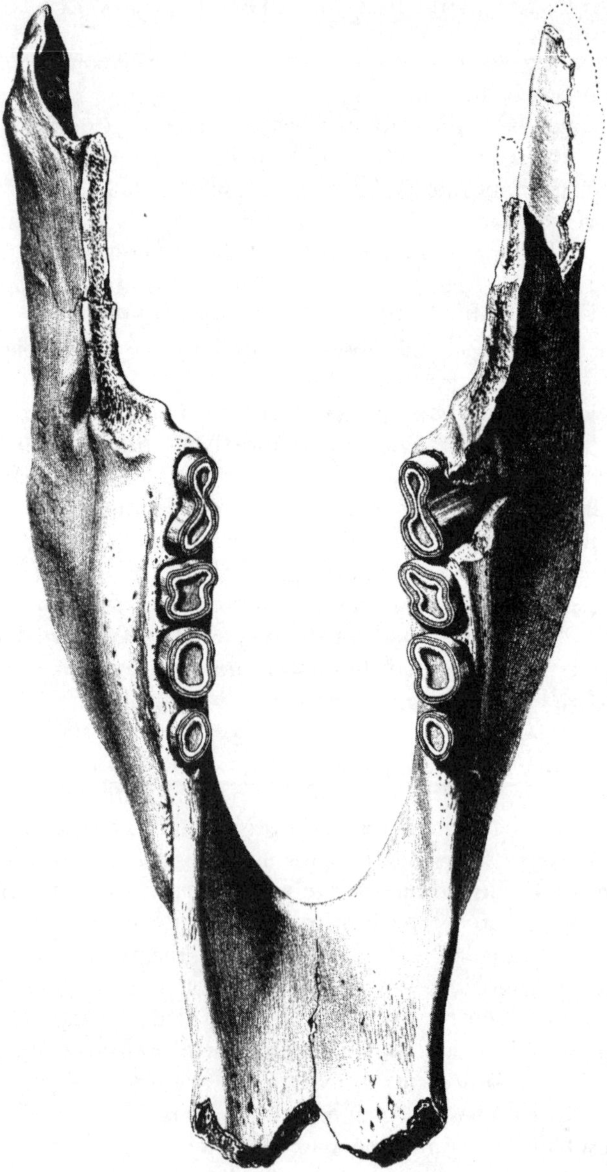

PLATE XVIII

Mylodon

The general form of the horizontal ramus of the jaw, is so well illustrated in the figures Pl. XVIII and XIX, that the description may be brief.

The symphysis is completely anchylosed, about four inches in length, and extended forward to the extremity of the jaw at a very slight angle with the inferior border of the ramus: it is of great breadth, smooth and gently concave internally, and suggests the idea of its adaptation for the support and gliding movements forwards and backwards of the free extremity of a long and well-developed tongue.

The exterior surface of the symphysis is characterized by the presence of two oval mammilloid processes, situated on each side of the middle-line, and about half way between the anterior and posterior extremes of the symphysis. A front view of these processes, of the natural size, is given in fig. 4, Pl. XIX: a side view of the one on the right side represented in the reduced figure.

Nearly four inches behind the anterior extremity of the above process is the large anterior opening of the dental canal: it is five lines in diameter, situated about one-third of the depth of the ramus of the jaw from the upper margin. The magnitude of this foramen, which gives passage to the nerve and artery of the lower lip, indicates that this part was of large size; and the two symphyseal processes, which probably were subservient to the attachment of large retractor muscles, denote the free and extensive motions of such a lip, as we have presumed to have existed from the size of the foramina destined for the transmission of its nervous and nutrient organs.

The angle of the jaw is produced backwards, and ends up in an obtuse point, slightly bent upwards; a foramen, one-third less than the anterior one, leads from near the commencement of the dental canal, to the outer surface of the jaw, a little below and behind the last molar tooth; this foramen presents the same size and relative position on both sides of the jaw. I find no indication of a corresponding foramen, or of symphyseal processes in the figures or descriptions of the lower jaw of the Megatherium, nor in the lower jaw of the Sloths, Ant-eaters, Armadillos, or Manises, which I have had the opportunity of examining with a view to this comparison. /

In the Megatherium the inferior contour of the lower jaw is peculiarly remarkable, as Cuvier has observed, for the convex prominence or enlargement which is developed downwards from its middle part. In the Mylodon the corresponding convexity exists in a very slight degree, not exceeding that which may be observed at the corresponding part of the lower jaw of the Ai, or Orycterope. A broad and shallow furrow extends along the outer side of the jaw, close to the alveolar margin, from the beginning of the coronoid process to the anterior dental foramen.

The base of the coronoid process begins external and posterior to the last grinder: the whole of the ascending ramus of the jaw, beneath the coronoid process is excavated on its inner side by a wide and deep concavity, bounded below by a well-marked ridge, which extends obliquely backwards from the posterior part of the alveolus of the last grinder to the inferior margin of the ascending ramus, which is bent inwards before it reaches the angle of the jaw.

The large foramen or entry to the dental canal is situated in the internal concavity of the ascending ramus of the jaw, two inches behind the last molar, three inches from the lower margin of the ramus, and nearly five inches from the elevated angle of the jaw: it measures nine lines in the vertical diameter, and its magnitude indicates the large size of the vessels which are destined to supply the materials for the constant renewal of the dental substance – a substance which from its texture must be supposed to have been subject to rapid abrasion. About an inch behind the dental foramen a deep vascular groove, about two lines in breadth, is continued downwards to the ridge which circumscribes the internal concavity of this part of the jaw, and perforates the ridge, which thus arches over the canal: this structure is present in both rami of the jaw. The mylo-hyoid ridge is distinctly marked about an inch and a half below the alveolar margin. Other muscular ridges and irregular eminences are present on the outer side of the base of the ascending ramus, and near the angle of the jaw; as shown in fig. 1, Pl. XIX.

From the preceding descriptions it will be seen that the lower jaw of the *Mylodon* is very different from that of the *Megatherium*; with that of the *Megalonyx* we have at present no means of comparing it. Among existing Edentata the Mylodon, in the form of the posterior part and angle of the jaw, holds an intermediate place between the Ai and the great Armadillo; in the form of the anchylosed symphysis of the lower jaw it resembles most closely the Unau or two-toed Sloth; but in the peculiar external configuration of the symphysis resulting from the mammilloid processes above described, the Mylodon presents a character which has not hitherto been observed in any other species of *Bruta*, either recent or fossil.

In conclusion it may be stated, that the teeth and bones here

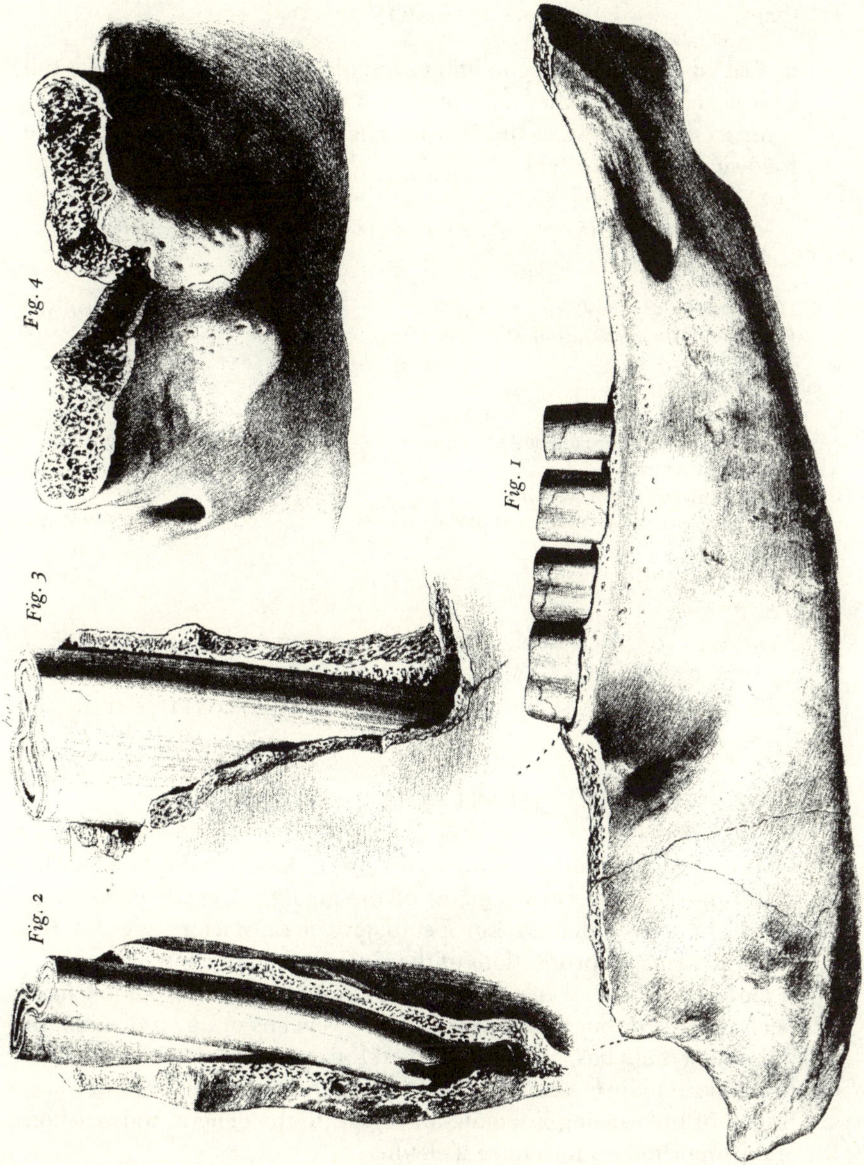

PLATE XIX

Mylodon

Fig. 1 is drawn 5/9 the size of Figs 2–4

described offer / all the conditions and appearances of those of a full grown animal; and that they present a marked difference of size as compared with those of the *Mylodon Harlani*, as will be evident by the following admeasurements.

ADMEASUREMENTS OF THE LOWER JAW OF MYLODON DARWINII

	Inches	*Lines*
Length (as far as complete)	17	6
Extreme width, from the outside of one ramus to that of the other	9	6
Depth of each ramus	4	9
Length of alveolar series	4	8
From first molar to broken end of symphysis	6	0
Breadth of symphysis	3	7
Longitudinal extent of symphysis	4	6
Circumference of narrowest part of each ramus	5	9

DESCRIPTION OF A CONSIDERABLE PART OF THE SKELETON OF A LARGE EDENTATE MAMMAL, ALLIED TO THE MEGATHERIUM AND ORYCTEROPUS, AND FOR WHICH IS PROPOSED THE NAME OF

SCELIDOTHERIUM[44] LEPTOCEPHALUM

Of the large Edentate quadrupeds that once existed in the New World, sufficient of the osseous remains of the gigantic Megatherium alone has been transmitted to Europe to give a satisfactory idea of the general form and proportions of the extinct animal.

Different bones of the Megalonyx, Mylodon, and Glyptodon have been described, but not sufficient of the remains of any individual of these subgenera has, hitherto, reached Europe, or been so described as to enable us to form a comparison between them and the Megatherium, or any of the existing Edentata, in regard to the general construction and proportions of the entire skeleton.

This state of our knowledge of the osteology of the singular giants of the Edentate Order renders the remains of the present animal

[44] Σκελις, *femur*; θηριον, *bellua*; in allusion to the disproportionate size of the thigh-bone.

peculiarly interesting, since, although the extremities are too imperfect to enable us to reconstruct the entire skeleton, a sufficient proportion of it has been preserved in the natural position to give a very satisfactory idea of its affinities to other Edentata, whose osteology is more completely known. /

The fossil remains here described were discovered by Mr Darwin in the same bed of partly consolidated gravel at Punta Alta, Northern Patagonia, as that in which the lower jaws of the *Toxodon* and *Mylodon* were embedded. The parts of the skeleton about to be described were discovered in their natural relative position, as represented at Pl. XX, indicating, Mr Darwin observes, that the sublittoral formation in which they had been originally deposited had been subject to little disturbance.[45] They include the cranium, nearly entire, with the teeth and part of the os hyoides; the seven cervical, eight of the dorsal, and five of the sacral vertebrae, the two scapulae, left humerus, radius and ulna, two carpal bones, and an ungueal phalanx; both femora, the proximal extremities of the left tibia and fibula, and the left astragalus.

The principal parts of the cranium which are deficient are the anterior extremities of both the upper and lower jaws, the os frontis, aethmoid bone, and the whole upper part of the facial division of the skull; but sufficient remains to show that the general form of the skull resembled an elongated, slender, sub-compressed cone, commencing behind by a flattened vertical base, slightly expanding to the zygomatic region, and thence gradually contracting in all its dimensions to the anterior extremity.

The Cape Ant-eater (*Orycteropus*), of all Edentata, most nearly resembles the present fossil in the form of its cranium, and next in this comparison the great Armadillo (*Dasypus gigas*, Cuv.) may be cited: on the supposition, therefore, that the correspondence with the above existing Edentals observable in the parts of the fossil cranium which do exist, was carried out through those which are defective, the length of the skull of the Scelidothere must have been not less than two feet. If now the reader will turn to Pl. XX he will see that this cranium is singularly small and slender in proportion to the rest of the skeleton, especially the bulky pelvis and femur, of which bones the latter has a length of seventeen inches, and a breadth of not less than nine inches; the astragalus, again, exceeds in bulk that of the largest Hippopotamus

[45] This beach is covered at spring tides; many parts of the skeleton were encrusted with recent *Flustrae*, and small marine shells were lodged in the crevices between the bones.

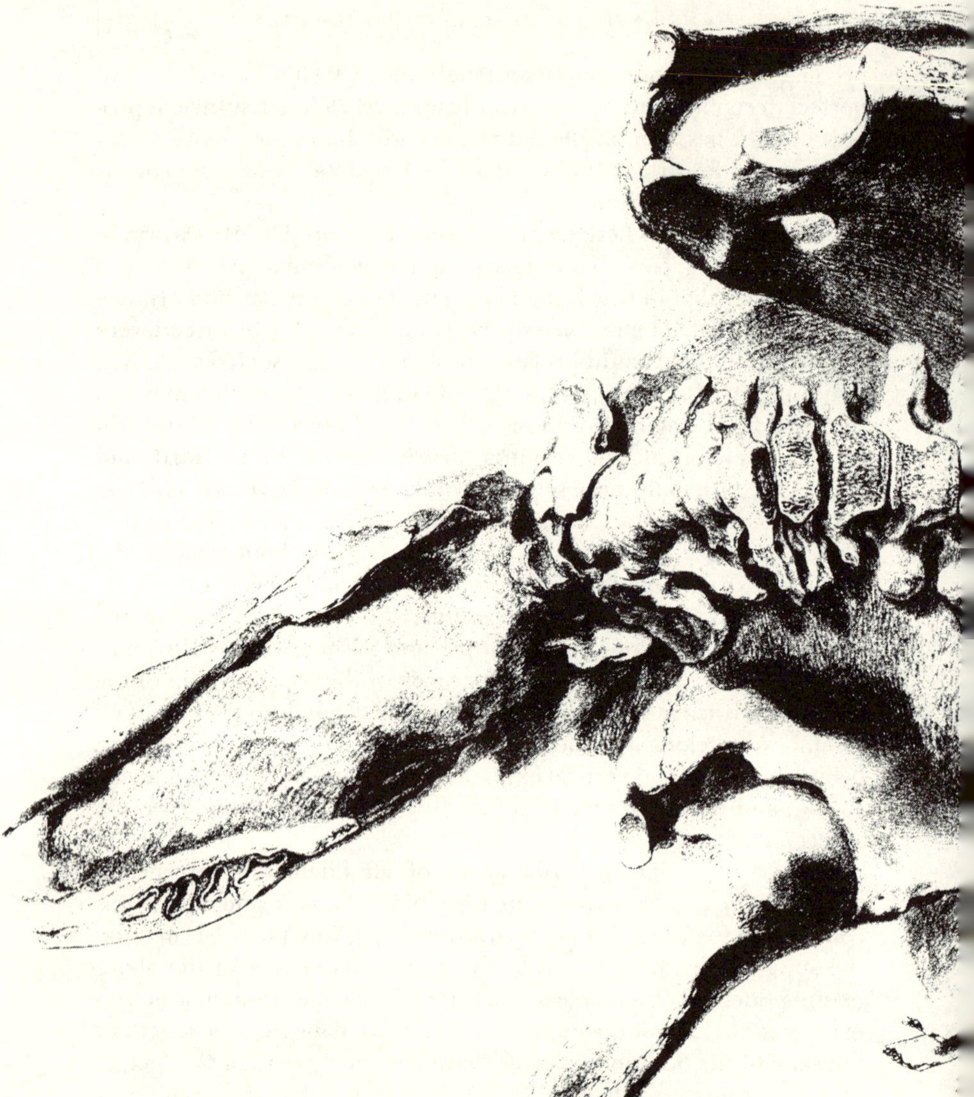

PLATE XX

Scelidotherium

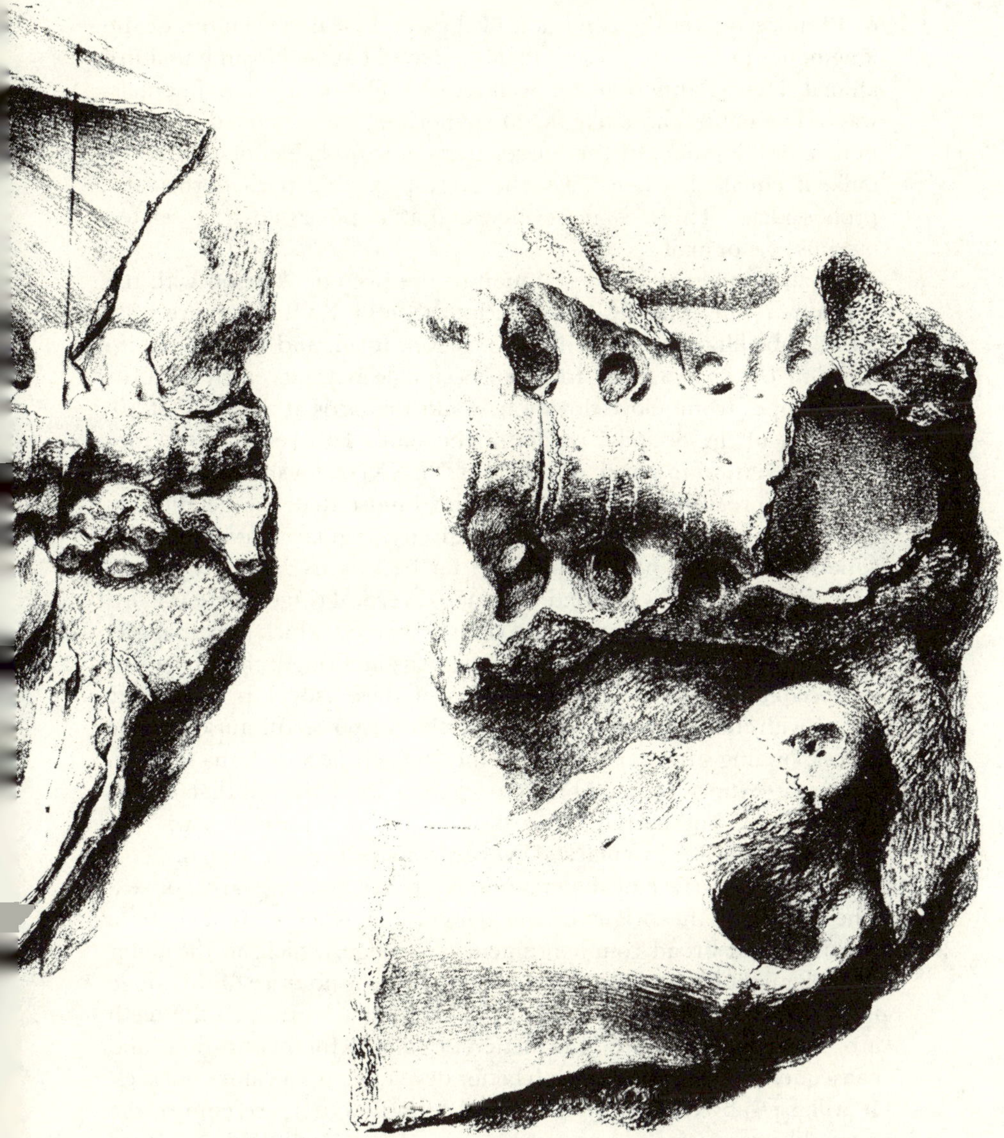

PLATE XX

Scelidotherium

or Rhinoceros; yet the condition of the epiphyseal extremities of the long bones proves the present fossils to have belonged to an immature animal. Hence, although the Scelidothere, like most other Edentals, was of low stature, and, like the Megatherium, presented a disproportionate development of the hinder parts, it is probable, that, bulk for bulk, it equalled, when alive, the largest existing pachyderms, not proboscidian. There is no evidence that it possessed a tesselated osseous coat of mail.

I shall commence the description of the present skeleton with the cranium. / The condyles of the occiput (See Pl. XXI, fig. 2) are wide apart, sub-elliptic, very similar in position, form, and relative size to those in *Orycteropus*. The foramen occipitale is transversely oval, its plane slopes from above downwards and forwards at an angle of 40° with that of the occipital region of the skull. This region, as before stated, is vertical in position (see fig. 1, Pl. XXI), of a sub-semicircular form, the breadth being nearly one-third more than the height; it is bounded above and laterally by a pretty regular curve; but the superior margin is not produced so far backwards as in *Orycteropus*. The occipital plane is bisected by a mesial vertical ridge; there is a less developed transverse curved intermuscular crest which runs parallel with and about half an inch below the marginal ridge: the surface of the occipital plane on the interspaces of these ridges is irregularly pitted with the impression of the insertion of powerful muscles. The corresponding surface is smooth in the Orycterope and Armadillos; in the great extinct Glossothere it resembles in character that of the Scelidothere; but in the forward slope of the occipital plane the Glossothere differs in a marked degree from the present animal.

The upper surface of the cranium is smooth and regularly convex. The extent of the origin of the temporal muscles is defined by a slightly-raised broad commencement of a ridge, which, in the older animal, might become more developed. There is no trace of this ridge in the Orycterope; but it exists in the Armadillos, in which the teeth are of a denser texture, and better organized for mastication, and consequently are associated with better developed masticatory muscles. It will be subsequently shown that the Scelidothere resembles the Armadillos in so far as it possesses a greater proportion of the dense ivory to the external caementum in its teeth, than does the Megatherium; while it differs widely from the Orycterope, in the structure of its teeth. The teeth, however, are fewer in the Scelidothere than in any Armadillo, and relatively smaller than in most of the species of

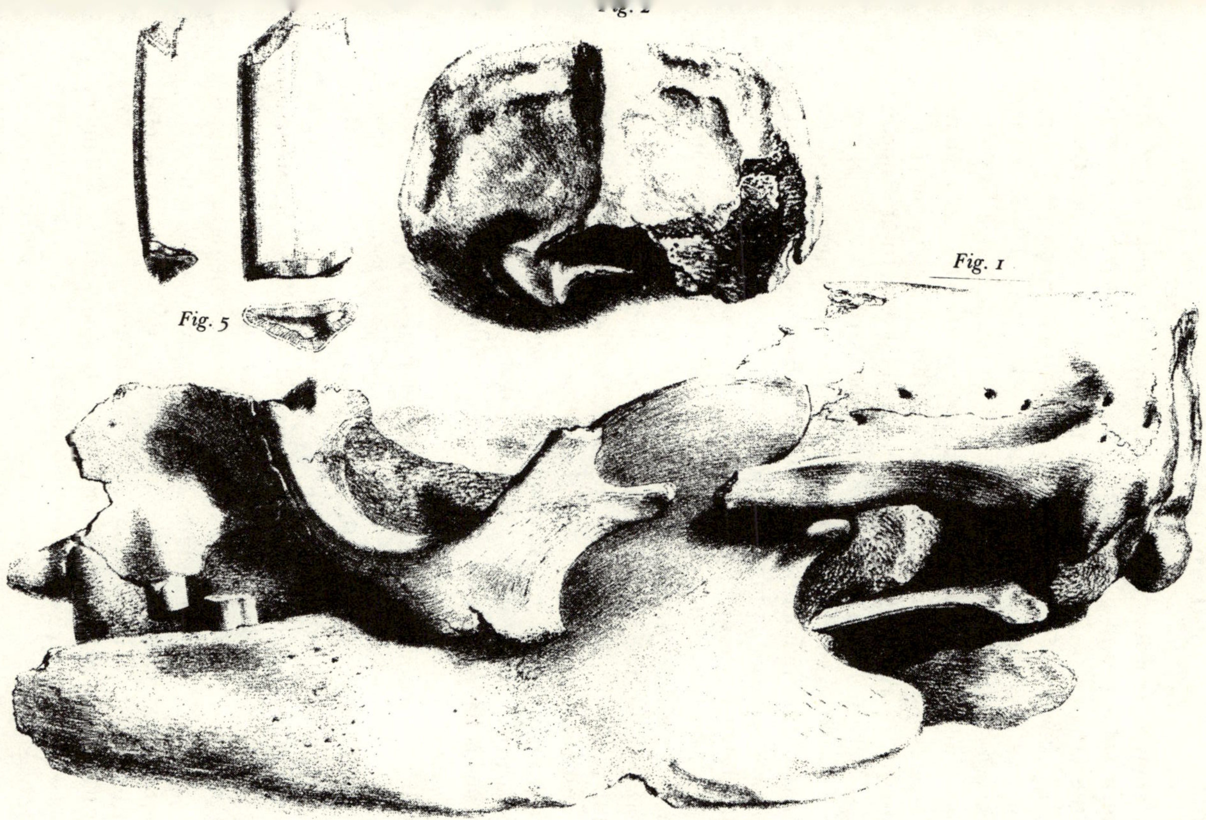

PLATE XXI

Scelidotherium

that family. Accordingly we find that the zygomatic arches are relatively weaker; and in this particular the *Scelidothere* corresponds with the Orycterope. The zygomatic process of the temporal commences posteriorly about an inch and a half from the occipital plane, its origin or base is extended forwards in a horizontal line fully four inches, where it terminates as usual in a thin concave edge, as shown on the right side in Pl. XXII. The free portion of the zygoma, continued forwards from the outer part of this edge, is a slender subcompressed process, half an inch in the longest or vertical diameter, and less than three lines in the transverse; the extremity of this process is broken off; the opposite extremity of the malar portion of the zygoma is entire, and obtusely rounded. The bony arch may have been completed by the extension of the temporal process to the malar one, but the two parts undoubtedly were not connected together by so extensive a surface as in / the Orycterope. On the other hand, if the zygomatic arch be naturally incomplete in the Scelidothere, the interspace between the malar and temporal portions must be relatively much less than in the Sloth or Ant-eater; for the broken end of the temporal part is separated from the obtusely rounded apex of the malar process in the present specimen by an interval of only one inch.

The articular surface (Pl. XXIII, fig. 2) beneath the zygoma for the lower jaw is flat and even, with the outer and inner margin slightly bent down, but having no definable anterior or posterior limits; its breadth is two inches. It differs from the corresponding surface in the Orycterope in being separated by a relatively wider interval from the tympanic bone, and in wanting consequently the support which the bony meatus auditorius gives in the Orycterope to the back part of the mandibular joint. The Armadillos differ still more from the Scelidothere in this important part of the cranial organization, inasmuch as the glenoid cavity is not only protected behind by the descending os tympanicum, but also in front by a corresponding vertical downward extension of the os malae. The Scelidothere in the general form and relative position of the surface for the articulation of the lower jaw resembles the Glossothere more closely than any other Edentate animal with which I have been able to compare it.

The malar bone of the Megatherium presents, as is well known, two characters, in which it conspicuously differs from that of the Orycterope and Armadillos, and approximates in an equally marked degree to the Sloths; these characters consist in a process ascending as if to complete the posterior circumference of the orbit, and another

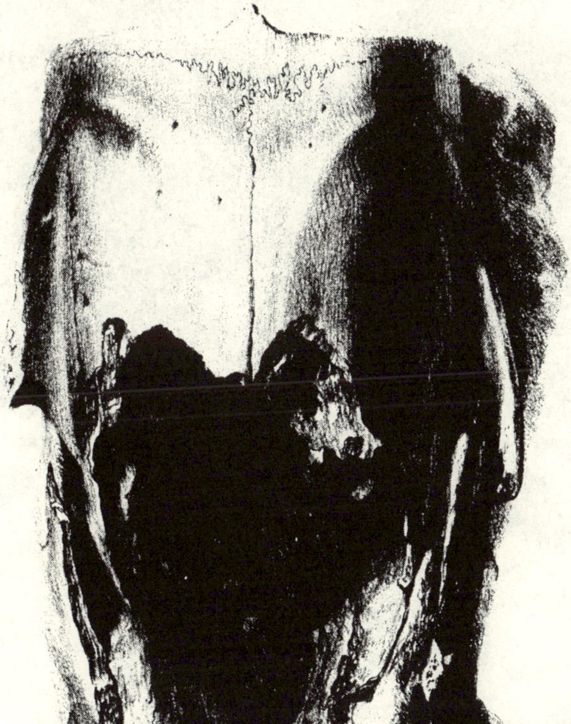

PLATE XXII
Scelidotherium

Fig. 1

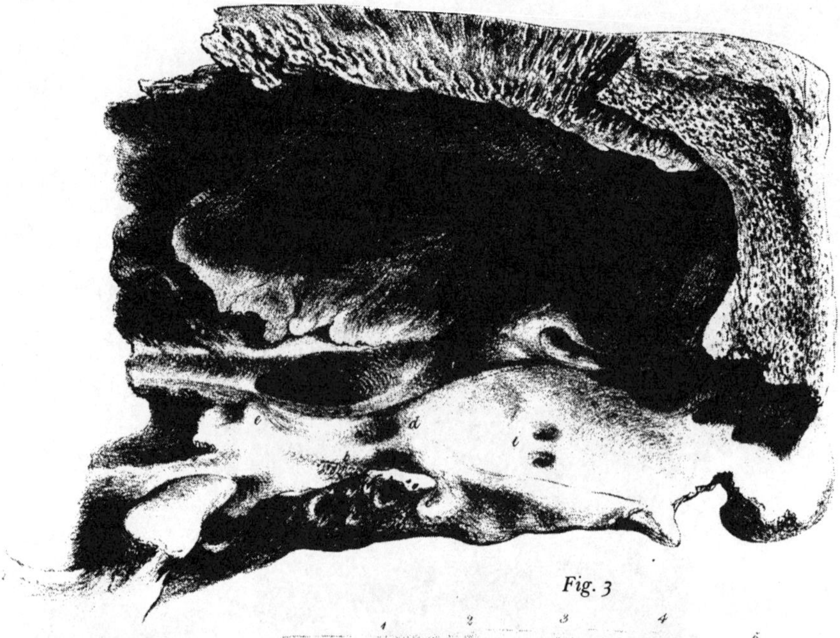

Fig. 2

Fig. 3

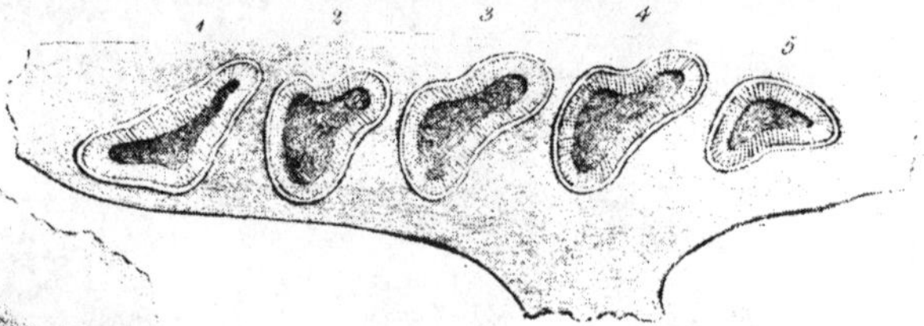

Fig. 4

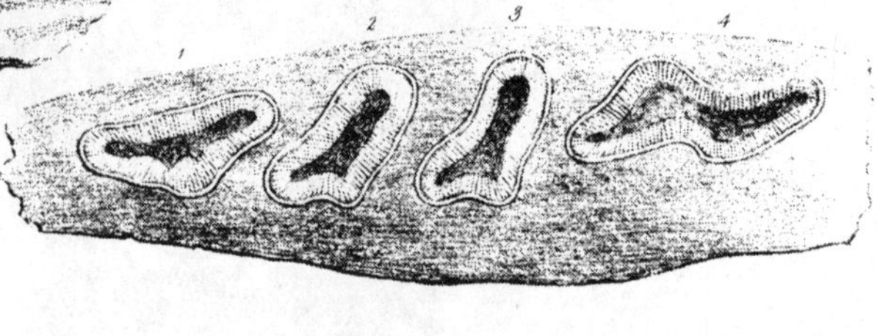

PLATE XXIII

Cranial cavity and dentition of Scelidotherium

process descending outside the lower jaw to give advantageous and augmented surface of attachment to the masseteric muscle, in its character of a protractor of the jaw. Now both these modifications of the malar bone are present in the Scelidothere, and are the chief if not the sole marks of the affinity to the Megatherium which the structure of the cranium affords. They are, however, the more interesting, perhaps, on that account, and because they are associated with other and more numerous characters approximating the species in question to the ordinary terrestrial as distinguished from the arboreal Edentata. For if the Scelidothere, instead of the Megathere, had been discovered half a century ago, and if its true nature and affinities had been in like manner elucidated by the genius and science of a Cuvier; and supposing on the other hand that the Megatherium instead of the Scelidothere had been one of the novel and interesting fruits of Mr Darwin's recent exploration of the coast of South America, then the affinities of the Megathere with the Sloths would undoubtedly have been viewed from a truer point than at the time when – the Scelidothere, and analogous transitional forms, being unknown – it was regarded as a gigantic Sloth.

Having indicated the principal characters of the cranium of the Scelidothere, / which determine its affinities among the *Edentata*, there next remains to be considered the relative position, extent, and connections, of the different bones composing the cranium.

The occipital bone constitutes the whole of the posterior, the usual proportion of the inferior, and a small part of the upper and lateral portions of the cranial cavity: there is a small descending ex-occipital process immediately exterior to the condyle: above this part the occipital bone is articulated to the mastoid process of the temporal, and the supra-occipital plate is joined by a complex dentated lambdoidal suture to the two parietals, without the intervention of interparietal or Wormian bones; the course and form of the lambdoidal suture is shown in Pl. XXII; it has the same relative position as in the Orycterope; in the Armadillos, the suture runs along the angle between the posterior and superior surfaces of the skull. The thickness of the occipital bone, at this angle, in the Scelidothere, exceeds an inch, and its texture consists of a close massive diploë, between the dense outer and inner tables (Pl. XXIII, fig. 1).

The squamous portion of the temporal bone has a very slight elevation, not extending upon the side of the cranium more than half an inch above the zygoma; it is thus relatively lower than in the

Orycteropus; but is similarly bounded above by an almost straight line (Pl. XXI, fig. 1). The mastoid process is small, compressed, with a rounded contour; immediately internal to it is a very deep depression, corresponding to that for the digastric muscle. But the most interesting features in this region of the temporal bone consist in the free condition of the tympanic bones, and the presence of a semicircular pit, immediately behind the tympanic bone for the articulation of the styloid element of the hyoid or tongue-bone: in these points we trace a most remarkable correspondence with the Glossothere, and in the separate tympanic bone the same affinity to the Orycteropus, as has been already noticed in the more bulky extinct Edental.

This correspondence naturally leads to a speculation as to the probable generic relationship between the Glossothere and Scelidothere: now it may first be remarked that the styloid articular depression is relatively much larger and much deeper in the Glossothere than in the Scelidothere; in the former its diameter equals, as we have seen, one inch; in the Scelidothere it measures only a third of an inch, the whole cranium being about two-fifths smaller; if we turn next to the anterior condyloid foramina, which in the Scelidothere are double on each side, we obtain from them evidence that the muscular nerve of the tongue could only have been one-third the size of that of the Glossothere. These proofs of the superior relative development of the tongue in the Glossothere indicate a difference of habits, and a modification, probably, of the structure of the locomotive extremities; and when we associate these deviations from the Scelidothere, with / the known difference in the position of the occipital plane, which in the Glossothere corresponds with that in the *Myrmecophaga* and *Bradypus*, we shall be justified in continuing to regard them, until evidence to the contrary be obtained, as belonging to distinct genera.

The parietal bones present an oblong regular quadrate figure, the sagittal suture running parallel with the squamous, and the frontal with the lambdoidal suture; there is scarcely any trace of denticulations in the sagittal suture; the bones are of remarkable thickness, varying, at this suture, from six to nine lines, and their opposed surfaces are locked together by narrow ridges, which slightly radiate from the lower to the upper part of the uniting surface: the substance of the bone consists of an uniform and pretty dense diploë; and there are no sinuses developed in it. We can hardly regard the extraordinary air-cells which occupy the interspace of the two tables of the skull in the parietal and occipital bones of the Glossothere (Pl. XVI, fig. 3) as a difference depending merely on age.

The frontal and aethmoid bones are broken away in the present cranium. The sphenoid commences two inches in front of the foramen occipitale; the fractured state of the skull does not allow its anterior or lateral limits to be accurately defined; its body is occupied with large air-sinuses; the only part, indeed, of this bone which is exposed to observation is that which forms part of the floor of the cranium; and this we shall now proceed to describe, in connection with the other peculiarities of the cranial cavity (fig. 1, Pl. XXIII). The body of the sphenoid is impressed on its cranial surface with a broad and shallow sella turcica (*a*), bounded by two grooves (*b b*), leading forwards and inwards from the carotid foramina (*c*); the line of suture between the sphenoid and occipital bones runs along a slight transverse elevation (*d*), which bounds the sella posteriorly; this suture is partially obliterated: a slight median protuberance (*e*) bounds the sella turcica anteriorly; there are neither anterior nor posterior clinoid processes. External to the carotid channel there is a wide groove (*f*) leading to the foramen ovale (*g*); this foramen is about one-third smaller than in the Glossothere, and therefore, as compared with the anterior condyloid foramina, indicates that the tongue was endowed with a greater proportion of sensitive than motive power in the Scelidothere: but in reasoning on the size of this nerve, it must be remembered that in both animals certain branches, both of the second and third divisions of the fifth pair of nerves, are to be associated with the persistence of large dental pulps, of which they regulate the secreting power. Anterior to the foramen ovale, and at the termination of the same large common groove, lodging the trunk of the fifth pair of nerves is the foramen rotundum (*h*); this leads to a very long canal, the diameter of which is five lines, being somewhat less than that for the third division of the fifth pair. The anterior sphenoid is broken away, so that no observation can be made on the optic foramina. /

The basilar process of the occipital bone is perforated at its middle by two small foramina (*i*) on the same transverse line, about half an inch apart.

In the Armadillo these foramina do not exist: in the Orycterope they are present, but open beneath an overhanging ridge, which is continued from them to the upper part of the anterior condyloid foramen on each side. The sella turcica of the Orycterope is deeper and narrower than in the Scelidothere; and is separated from the basilar occipital process by a transverse ridge, which sends forward two short clinoid processes; two smaller anterior clinoid processes project backwards from the angle of the anterior boundary of the sella turcica. The foramina ovalia and rotunda open in the same continuous groove, as in the Glossothere and Scelidothere, but they are relatively wider apart; and the canal for the third division of the fifth pair is shorter, and runs more directly outwards.

The petrous bone in the Scelidothere is relatively larger than in the Glossothere, but this probably arises from the precocious development of the organ of hearing in the present immature specimen in obedience to the general law. The trunk of the fifth pair of nerves does not impress it with so deep and well defined a groove as in the Glossothere; the elliptic internal auditory foramen (*k*) is situated about the middle of the posterior surface; behind this is the aqueductus vestibuli; and immediately posterior to the petrous bone is the foramen jugulare (*l*): the shape of the os petrosum agrees more with that of the Armadillo than with that of the Orycterope. An accidental fracture of the right os petrosum demonstrates its usual dense and brittle texture, and at the same time has exposed the cochlea with part of its delicate and beautiful lamina spiralis. The conservation of parts of the organs of vision in certain fossils, has given rise to arguments which prove that the laws of light were the same at remote epochs of the earth's history as now; and the structures I have just mentioned, in like manner, demonstrate that the laws of acoustics have not changed, and that the extinct giants of a former race of quadrupeds were endowed with the same exquisite mechanism for appreciating the vibrations of sound as their existing congeners enjoy at the present day.

The brain, being regulated in its development by laws analogous to those which govern the early perfection of the organ of hearing, appears to have been relatively larger in the Scelidothere than in the Glossothere: it was certainly relatively longer; the fractured cranium gives us six inches of the antero-posterior diameter of the brain, but the analogy of the Orycterope would lead to the inference that it extended further into the part which is broken away. The greatest transverse diameter of the cranial cavity is four inches eight lines: these dimensions, however, are sufficient to show that the brain was of very small relative size in the Scelidothere; and, both in this respect, and in the relative position of its principal masses, the brain of the extinct Edental closely accords with the general character of this organ in the existing species of the same Order. We perceive by the obtuse / ridge continued obliquely upwards from above the upper edge of the petrous bone, that the cerebellum has been situated wholly behind the cerebrum; we learn also from the same structure of the enduring parts that these perishable masses were not divided, as in the Manis, by a bony septum, but by a membraneous tentorium, as in the Glossothere and Armadillos: in the Orycteropus, as had been before remarked, there is a strong, sharp, bony ridge extending into each side of the tentorium. The vertical diameter of the cerebellum and medulla oblongata equals that of the cerebrum, and is two inches three lines: the transverse diameter of the cerebellum was about three inches nine lines; its antero-posterior extent about one inch and a half. The sculpturing of the internal surface of the cranial cavity bespeaks the high vascularity of the soft parts which it contained, and there are evident indications that the upper and lateral surfaces of the brain had been disposed in a few simple parallel longitudinal convolutions. The two anterior condyloid foramina (*m*) have the same relative position as the single corresponding foramen in the Glossothere, Orycterope, and Armadillos, and the inner surface of the skull slopes outwards from these foramina to the inner margin of the occipital condyle.

Of the bones of the face there remain only portions of the malar, lachrymal, palatine, and maxillaries. The chief peculiarities of the malar bone have been already noticed: the breadth of the base of the descending masseteric processes is two inches two lines; its termination is broken off: the length of the ascending post-orbital process of the malar cannot be determined from the same cause, but it is fortunate that sufficient of this part of the cranium should have been preserved to give this evidence of the affinities of the Scelidothere to the Megathere. The malar bone is continued anteriorly, in a regular curve forwards and upwards, to the lachrymal bone, and completes, with it, the anterior boundary of the orbit: the size of the orbit is relatively smaller than in the Orycterope, and still less than in the Ant-eaters: here, however, we have merely an exemplification of the general law which regulates the relative size of the eye to the body in the mammalia. The malar bone does not extend so far forwards in front of the orbit as in either the Orycterope or Armadillo; in the inclination, however, with which the sides of the face converge forwards from the orbits, the Scelidothere holds an intermediate place between the Armadillos and Orycterope.

The lachrymal bone does not extend so far upon the face in the

Scelidothere as in the Orycterope; in which respect the Scelidothere resembles more the Megathere. The foramen for the exit of the infra-orbital nerve has the same situation near the orbit as in the Megathere; its absolute distance from the anterior border of the orbit is only half that in the Orycterope. The foramen is single in the Scelidothere, as in the Orycterope; in the Megathere there are two or three antorbital foramina. The vertical diameter of this foramen is eight lines, the transverse diameter four lines. So much of the outer surface of the superior maxillary bones as has been preserved, / is smooth and vertical. Each superior maxillary bone contains the sockets of five teeth, occupying an antero-posterior extent of three inches seven lines (Pl. XXII and XXIII, fig. 3). The posterior alveolus is situated just behind the transverse line, extending across the anterior boundary of the orbits; the remaining sockets of the molar series extend forwards three inches in front of the orbits. In the Megatherium, the roots of the five superior molars are all situated behind the anterior boundary of the orbit: in the Orycteropus, on the contrary, the grinders are all placed in advance of the orbit; so that the Scelidothere resembles that species more than the Megathere in the relative location of the teeth. The palatal interspace between the roots of the last molar tooth of each series is eleven lines; the palate gradually though slightly widens, as it advances forwards: the posterior margin of the palate is terminated by an acute-angled notch. In the breadth of the bony palate the Scelidothere is intermediate between the Megathere and Orycterope.

The anterior of the upper molars is represented at figs 3, 4, and 5, Pl. XXI, and at 1, fig. 3, Pl. XXIII; it corresponds closely in form and size with the opposite molar below; the base of the triangle given by its transverse section is turned inwards and obliquely forwards.

The second molar of the upper jaw, also presents in transverse section a triangular form, with the angles rounded off; but the inner side of the tooth is traversed by a longitudinal groove. The largest diameter of the transverse section, which is placed obliquely as regards the axis of the skull, measures ten lines and a half; the opposite diameter of the tooth is six lines.

The third and fourth molars present the same form and size, and relative position as the second.

The fifth molar is the smallest of the series; its transverse section gives an inequilateral triangle, with the corners rounded off; the broadest side is turned outwards, and is slightly concave; the antero-posterior diameter of this tooth is seven lines; the transverse four lines. The length of the teeth in the upper jaw is about two inches and a half.

It is almost superfluous to observe that the teeth of the Scelidothere, as in other *Bruta* are without fangs, and have their inserted base excavated by large conical cavities, for the lodgment of a persistent pulp. The tooth is composed of a small central body of coarse ivory or 'dentine', traversed by medullary canals, which at the periphery of the coarse dentine anastomose by loops, from the convexity of which the calcigerous tubes are given off which form the fine dentine: the layer of this substance, which immediately surrounds the coarse dentine, is about one line and a half in thickness, and the whole is invested with a very thin coating of cement. The teeth of the Scelidothere thus / present a more resisting structure than do those of the Mylodon; having a larger proportion of the dense ivory composed of the minute calcigerous tubes, and a much smaller proportion of the softer external caementum; in this respect the Scelidothere recedes farther from Megathere, and approaches nearer the Armadillos than does the Mylodon.

The lower jaw resembles, in the general form of the posterior moiety which is here preserved, that of the Sloth and Mylodon more than that of any other Edentate species. Its deep posterior angle is produced backwards, and a broad coronoid process rises and nearly fills the zygomatic space; the condyle is flat, as the glenoid surface has already indicated; its transverse diameter is an inch and eight lines; its antero-posterior diameter seven lines: it is principally extended inwards beyond the vertical line of the ascending ramus. The lower contour of the jaw describes an undulating line; which, commencing from the posterior angle, is at first gently convex, then slightly concave, then again convex, below the alveoli of the teeth, where it is rounded and expanded, as in the Orycterope. The fractured condition of the right ramus of this part fortunately exposed the roots of the four grinding teeth, which constitute the dental series on each side of the lower jaw. The length of the jaw occupied by these four alveoli is three inches ten lines, which exceeds a little that of the opposed five grinders above; the ramus of the jaw gradually diminishes in all its dimensions anterior to the molar teeth; the dental canal passes in a gentle curve below, and on the inner side of the alveoli, whence it gradually inclines to the outer wall of the jaw.

The whole ascending ramus of the jaw consists of a very thin plate of bone; it is slightly concave on the inner side, and the inferior margin of the produced angle inclines inwards, as in the Mylodon and Sloth; it is impressed on the outer side with two shallow depressions, and two parallel ridges, both following the gentle curvature of the part. There is a foramen on the outer side of the ramus at the anterior part of the base of the coronoid process corresponding with that in the lower jaw of the Mylodon, but the longitudinal channel which runs along the outer side of the alveolar processes is wanting, and the expansion at the base of those processes is more sudden and relatively greater; the general correspondence, however, between these lower jaws is such as would lead to the idea that they belonged to animals of the same genus, were it not that the teeth present modifications of form in the Scelidothere, as distinct from those of the Mylodon, as are any of the minor dental differences on which genera or sub-genera of existing Mammalia are founded in the present state of Zoological Classification.

To make this distinction more readily intelligible, I have given a view of the transverse section of the teeth in the right ramus of the lower jaw (fig. 4, Pl. XXIII), corresponding with that of the *Mylodon Darwinii* (Pl. XVII, fig. 5). In the present / sub-genus the antero-posterior extent of the four alveoli of the lower jaw nearly equals four inches, and is relatively greater than in the Mylodon, although the teeth are placed closer together; this is owing to their greater relative size. The first molar tooth presents the simplest form; its transverse section is a compressed inequilateral triangle with the angles rounded off; the longest diameter of this section which is parallel with the inner alveolar border is eleven lines, the transverse diameter almost six lines; the base or broadest side of the triangle is turned inwards, and is slightly concave; the two smaller sides are also slightly concave.

The second molar is placed more obliquely in the jaw; the long axis of its transverse section intersects at an acute angle that of the jaw itself; the transverse section presents a compressed or oblong form, with the larger end next the outer side, and the smaller end next the inner side of the jaw; this end is simply rounded, but the outer end presents a sinuosity, corresponding to a broad groove which traverses the whole length of the outer side of the tooth; the anterior, which corresponds to the internal side or base of the transverse section of the preceding molar, is slighly concave.

The third molar has nearly the same form and relative position as the preceding; the long diameter of the transverse section is, in both, ten lines and a half; the principal transverse diameter is, in the second molar five lines, in the third nearly six; the difference of form observable in these as compared with the two middle grinders of the Mylodon is well marked; in the latter these teeth are impressed with a longitudinal groove on their inner sides; in the Scelidothere they have a similar impression along their outer but not along the inner side.

In the last molar the resemblance is much closer, and the modification of form by which it differs from the preceding ones is of the same kind; the transverse section gives an irregular oblong figure with its axis nearly parallel with that of the jaw, and constricted at the middle by sinuosities produced by two wide channels which traverse longitudinally, one the outer, the other the inner side of the tooth; the latter groove is much wider and shallower in the Scelidothere than in the Mylodon. The two lobes produced by these grooves are more equal in Scelidothere; the anterior one is concave on its anterior surface instead of convex as in the Mylodon; the posterior one is more compressed; the longitudinal or antero-posterior diameter of the transverse section of this tooth is one inch five lines; the greatest transverse diameter is nine lines; the diameter of the isthmus joining the lobes is three lines and a half; the entire length of this tooth is three inches three lines.[46] /

VERTEBRAL COLUMN

Of this part of the skeleton of the Scelidothere, Mr Darwin's specimen includes, as is represented in Plate XX, the cervical, part of the dorsal, and the sacral series of vertebrae in a more or less perfect condition.

The cervical vertebrae present the ordinary mammalian number, seven, and are free, or so articulated as to have permitted reciprocal movement upon each other. Their transverse processes are perforated as usual for the vertebral arteries. These processes in the atlas are remarkable for their great breadth, length, and thickness; and indicate the muscular forces which must have worked the head upon the spine to have been very powerful. The axis is provided with a robust 'processus dentatus', having a base equal in breadth to the body of the axis itself; and a smooth articular convexity on the side of the apex on which the ring of the atlas rotated. The line of union between the axis and its characteristic process, which here resembles the body of an

[46] It requires little stretch of imagination to conceive that this more complex posterior tooth (Pl. XXIII, fig. 4, 4) in the lower jaw is the representative of the two smaller posterior teeth (ib. fig. 3, 4, and 5) of the upper jaw conjoined.

abortive vertebra, is very distinct. The transverse processes of the vertebra dentata are comparatively feeble, but this condition is amply compensated for by the great development of the spinous process. (Pl. XXIV, fig. 1.) This process is bent backwards at nearly a right angle, overlaps with its reflected extremity the spine of the third cervical vertebra, and rests by its base, on the under part of which are the posterior articular surfaces, upon the broad and strong anterior oblique processes of the third vertebra.

The third, fourth, fifth, and sixth cervical vertebrae have moderately developed and pointed spinous processes: their transverse processes are broad, and extend obliquely backwards, and slightly overlap each other. On the under part of the transverse process of the sixth cervical vertebra there is the fractured base of what I conjecture to have been an expanded aliform plate, analogous to that observable in the corresponding vertebra of the Orycterope. The seventh cervical vertebra has part of the articular depression for the head of the first rib upon each side of its body: the transverse process is feebly developed, but the spine is double the height and size of those of the preceding vertebrae.

The spinous process of the first dorsal vertebra in like manner rises to twice the height of the preceding spine of the seventh cervical, and preserves an equal antero-posterior diameter from its base to its summit, which is thick and slightly bent backwards: four or five succeeding dorsal vertebrae give evidence of having been surmounted by spines of equal height and strength. The transverse processes of these dorsal vertebrae present bold concavities on their inferior part for the reception of the tubercles of the ribs, and they gradually ascend upon the base of the spines as the vertebrae are placed further back, so as to increase the expansiveness / of the chest. The state of the fossil did not afford further information as to the condition of this part of the vertebral column, but the parts which have been preserved are precisely those from which the most interesting inferences as to the affinities and habits of the extinct quadruped can be deduced.

Whether the Megatherium be most nearly allied to the tribes of the Sloth or Armadillo has been a question under recent discussion, and, as a corollary of this problem, whether its habits were those of a scansorial or of a fossorial quadruped. For, strange as it may appear at first sight, there have not been wanting arguments, and those urged by an anatomist to whom we owe much novel and interesting information respecting the extict Edentata, in support of the belief that the

Fig. 1

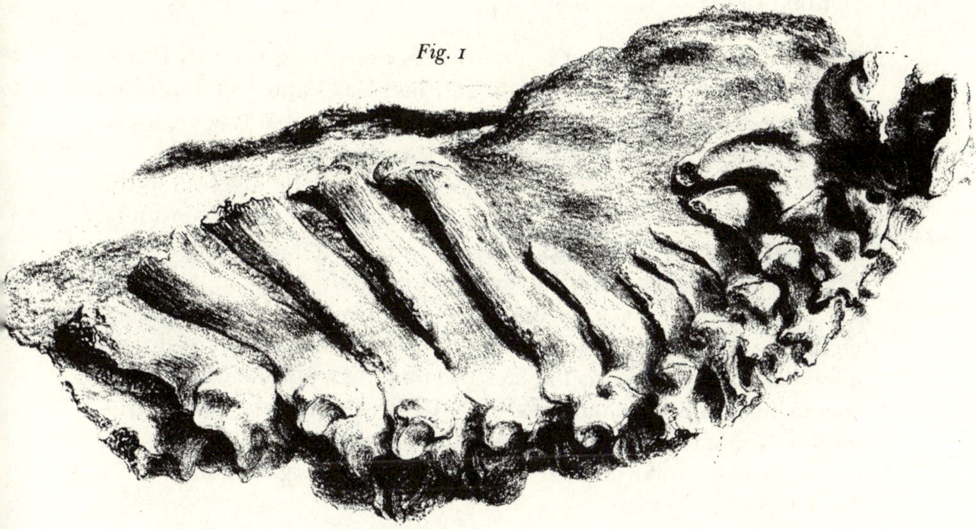

Fig. 2

Fig. 3

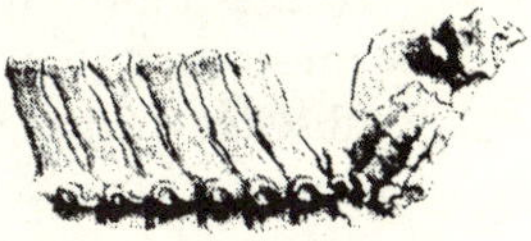

Fig. 4

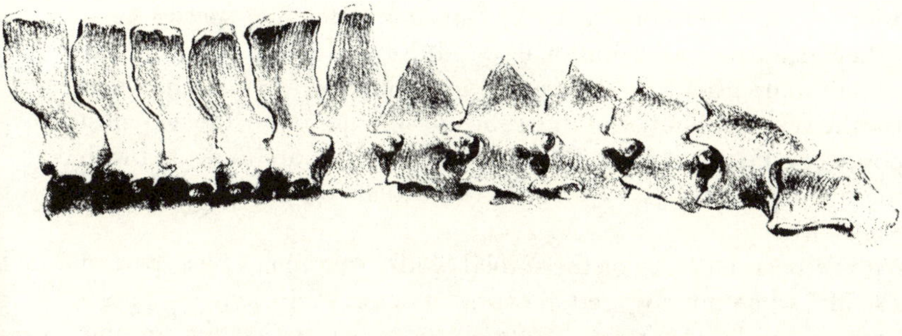

PLATE XXIV

Cervical and anterior dorsal vertebrae

Fig. 1. Scelidothere *Fig. 2. Orycterope* *Fig. 3. Armadillo* *Fig. 4. Great Anteater*

Megatherium, gigantic and ponderous as must have been its frame, actually climbed trees like a Sloth, and had claws and feet organized for prehensile actions, and not in accordance with that type by which they are usually adapted for digging up the soil.[47]

Now, in whatever degree the Megatherium may be involved in this question, the smaller Megatherioid species at present under consideration must be at least equally implicated in it. In the adaptation of the frame of a mammiferous quadruped for especial and peculiar actions and modes of life, such as for climbing and living in trees, or for burrowing and seeking concealment in the earth, not only the immediate instruments, as the feet, are modified, but the whole of the osseous and muscular fabric is more or less impressed with corresponding adaptations, whilst at the same time these special adjustments are invariably subordinated to the type of organization which characterizes the group.

The type of the order *Bruta* or *Edentata* is well-marked; one or more claws of unusual length and strength, characterize the fore-feet and sometimes the hind-feet in every genus, and the term 'Macronykia' would more aptly designate them than the term which Cuvier substituted for the good old Linnaean appellation. The uniform absence of true roots to the teeth, where these are present, is another general character; the skeleton exhibits many well-marked peculiarities common to the whole order; while at the same time it is modified in various modes and degrees in accordance with the peculiar habits and exigencies of the species.

One of the regions of the skeleton which manifests adaptive modifications of this kind in the most remarkable degree is the cervical division of the vertebral column. In one edentate species it is lengthened out by two additional vertebrae more than in any other mammal; in another it is reduced by anchylosis to as great an extent below the regular number of moveable pieces: and these, the two most opposite conditions of the cervical vertebrae which are to be met with in the mammiferous class are related to equally diverse and opposite habits of life. /

With respect to the *Ai*, or three-toed Sloth, 'an animal, great part of whose life, when not engaged in eating, is spent in sleeping on trees – an easy attitude for repose is most essential to its comfortable

[47] Lund, *Videnskabernes Selskabs, Natur: og Mathem. Afhandlinger*, Kiöbenhavn, vol. viii.

existence; and accordingly we find, that the auxiliary vertebrae at the base of the neck contribute to produce that flexibility of this organ which allows the head of the animal to incline forwards and rest upon its bosom.' Dr Buckland, from whose Paper on the 'Adaptation of the Structure of the Sloths to their peculiar Mode of Life',[48] the preceding judicious physiological remark is quoted, adduces the authority of Mr Burchell in proof that the Sloth can in a remarkable manner and with great facility twist its head quite round, and look in the face of a person standing directly behind it, while at the same time the body and limbs remain unmoved. A single glance at the length and slenderness of the cervical region of the spine, and of the feeble condition of the transverse and spinous processes in the vertebrae composing that part of the skeleton of the Sloth, is enough to show its adaptation to increase the rotatory motion and flexibility of the neck.

In describing the skeleton of a species of Armadillo (*Dasypus 6-cinctus*, Linn.)[49] I was led in like manner to point out the subserviency of the peculiarities of the cervical vertebrae to the habits and mode of life of that animal; observing that the 'anchylosis of the cervical vertebrae obtains in the *Cetacea*, as well as in the genus *Dasypus*, and that as in the aquatic order this firm connection of the cervical vertebrae assists materially in enabling the head to overcome the resistance of the dense fluid through which they perpetually move, so in the Armadillos a like advantage may be derived from this structure during the act of displacing the denser material in which they excavate their retreats'.[50]

Having in view these well-marked examples of the subserviency of the structure of the bones of the neck to the habits of existing species of the order *Bruta*, I proceeded to investigate the structure of the corresponding part of the skeleton in the *Scelidotherium*, hoping thereby to gain a new and useful element in the determination of the problem at present under discussion, as to the affinities and habits of the extinct Megatherioid quadrupeds.

The fossil, in its original state, yielded a view of so much of the anterior part of the bodies of the cervical vertebrae as proved that they

48 *Linn. Trans.*, vol. xvii (1833), p. 17.

49 *Zool. Proceedings*, 1832, p. 134.

50 The anterior prolongation of the sternum in front of the neck and the corresponding anterior position of the clavicles and scapulae occasions a transference of such a proportion of the moving powers of the head from the cervical vertebrae to these bones in the mole, as renders any modifications of these vertebrae, like those in the Armadillo, uncalled for.

were neither so numerous as in the Sloth, nor anchylosed together as in the Armadillos: after a long and careful chiselling at the hard matrix in which they were imbedded, the transverse / and spinous processes were exposed to view, as they are represented in Plates XX and XXIV. The description of these processes has already been given.

On comparing the cervical vertebrae of the Scelidotherium with those of the existing *Bruta*, the closest resemblance to them was found in the skeleton of the Orycterope. Now this quadruped, though not so rapid a burrower, or so strictly a subterranean species as the Armadillos, participates, nevertheless, to a certain extent, in their fossorial habits, and is closely allied to them in general structure: it differs from them, indeed, mainly in a modification of the dental system, in the absence of dermal armour, and of anchylosis of the cervical vertebrae. But the advantages which, as a burrower, it would have derived from the latter structure, are compensated for by the shortness of the cervical vertebrae, and by the great development and imbricated or interlocking co-adaptation of the transverse and anterior spinous processes of the cervical vertebrae. The analogous quadruped in the South American Continent – the great ant-eater (*myrmecophaga jubata*) which uses its powerful compressed fossorial claws for breaking through the hard walls of the habitations of its insect prey, but which does not excavate a subterraneous retreat for itself, presents the cervical vertebrae of a more elongated form, and without that development of the spinous and transverse processes which tend to fix the neck and increase the size of the muscles which move the head: and, if we could conceive that its fore-feet were employed to scratch up vegetable roots, instead of disinterring termites, there would be no reason to expect any modification of the cervical vertebrae as a direct consequence of such a difference in the application of its fossorial extremities: when, therefore, we find that the cervical vertebrae do actually differ in two myrmecophagous species, to the extent observable in the Cape and South American ant-eaters, we arrive legitimately at the conclusion that such difference relates to fossorial habits of the one species, in which habits the other does not participate.

Now, therefore, if this conclusion be just in regard to the Orycterope, it must bear with more force upon the question of the habits of the Scelidotherium as the mechanism for strengthening the connection of cervical vertebrae, and for augmenting the surface of attachment of the muscles which worked the head and neck, is more strongly wrought out in that extinct species.

The great size and strength of the spinous process of the dentata, and the mode in which it is interlocked with the spinous and oblique processes of the third cervical, together with the imbricated disposition of the transverse processes of this and the succeeding vertebrae, and the remarkable height of the dorsal spines, all combine to indicate in a very striking manner, if not to demonstrate, that the conical head of the present species, which is comparatively small and slender, and / for its own mere support requiring therefore no such mechanism, was used in aid of the fossorial actions of the extremities.

As the cervical vertebrae of the Megatherium have their processes comparatively weaker than in the Scelidotherium, and the anterior dorsal spines are relatively shorter, it may be concluded, that whatever were the extent or nature of the fossorial labours of the enormous claws with which it was provided, the head did not co-operate with the digging implements in their especial task in the same degree as in the Scelidothere and Orycterope. At the same time there is no modification of the cervical region of the spine of the Megathere corresponding with those which we have seen to be subservient to the arboreal habits of the sloth, a remark which will not be deemed superfluous by those who have perused the acute observations and arguments adduced by M. Lund in favour of the scansorial character of the extremities of the Megatherium and Megalonyx.

The fragments of the dorsal vertebrae and ribs of the Scelidotherium, which are figured in Plate XX, offer no modifications which need detain our attention; they closely conform, excepting in the greater relative height of the anterior dorsal spines, already noticed, with the Megatherioid type. The sacrum manifests in its vast expanse, the great development of the posterior transverse processes to join the ischium, the capacious medullary cavity, and wide nervous foramina, a like conformity with the Megatherium, and a corresponding harmony with the disproportionate bulk of the hind legs.

BONES OF THE EXTREMITIES

The Scapula in its double spine, the osseous arch formed by the confluence of the acromion with the coracoid process, and the substitution of a distinct foramen for the suprascapular notch, agrees with that of the Megatherium: but the span of the acromial arch is relatively wider, and the surface for the articulation of the clavicle is better marked. This articular surface, which is distinctly shewn upon the acromion of both the scapulae in Pl. XX is the more interesting, as being the only evidence of the clavicle of the Scelidothere which we at present possess; but it is enough to prove that this quadruped enjoyed all the advantages in the actions of the fore-extremity, which arise out of the additional fixation of the shoulder-joint afforded by the clavicle – a bone which the extinct Megatherioids are the largest of the mammiferous class to possess in a completely developed state. The form, position, and aspect of the glenoid cavity for the humerus closely correspond with the condition of the same part in the Megatherium. The limits of the acromial and coronoid portions of the arch were still defineable in the / present skeleton, which indicates the nonage of the individual in the unanchylosed condition of most of the epiphyses.

In regard to the presence of a clavicle in the Megalonyx M. Lund has deduced certain conclusions, which, if well founded, would be equally applicable to the present allied species, and to the great Megatherium. I am induced, therefore, to offer a few physiological observations on that bone, which appear to me to lead to a more correct interpretation of its uses and relations in the great mammiferous animals now under consideration.

When the anterior extremities in mammalia are used simply for the purpose of progressive motion on dry land, as in the Pachyderms and Ruminants, or in water, as in the Cetaceans, there is no clavicle; this bone is introduced between the sternum and acromion, in order to give firmness and fixity to the shoulder-joint when the fore-leg is to discharge some other office than that of locomotion. In these cases, however, the clavicle exists in various degrees of development, and even its rudiment may be dispensed with in some of the actions which require a considerable extent of lateral or outward motion, and of freedom of rotation of the fore-limb. When, therefore, we find the clavicle fully developed in the skeleton of an extinct mammiferous animal, and so placed as to give the humeral articulation all the benefit of this additional mechanism, we may confidently expect that it will afford an insight into the habits and mode of life of such extinct species. M. Lund[51] has argued from the clavicle of the Megalonyx, that it climbed like a Sloth. 'Animals,' says Sir C. Bell,[52] 'which fly or dig, or climb, as Bats, Moles, Porcupines, Squirrels, Ant-eaters, Armadilloes,

[51] Loc. cit.
[52] *Bridgewater Treatise*, p. 46.

and Sloths, have this bone; for in them, a lateral or outward motion is required.' But in regard to the present problem, we have to enquire whether the clavicle manifests any modifications of form, of strength, or development in relation to the special differences of these several actions, with which its presence is asserted to be associated?

In mammals which fly, the clavicle is always complete: the rabbit, the fox, and the badger are instances of burrowing animals in which the clavicle is absent or rudimental. The presence of a perfect clavicle is not more constant in climbing quadrupeds. The Ai, for example, has an incomplete clavicle, which is attached to the acromion process, and terminates in a point about one-fourth of the distance between the acromion and the top of the sternum, to which the clavicular style is attached by a long slender ligament: the advantage, therefore, which a perfect clavicle affords in the fixation of the shoulder-joint, is lost to this climber *par excellence*. Again, the Bears, which are the bulkiest quadrupeds that are gifted with the faculty of climbing, and this in so perfect a degree that the Sun-bears of the Eastern Tropics may be termed arboreal animals – these scansorial / quadrupeds are destitute of even the smallest rudiment of a clavicle, as I have ascertained by repeated careful dissection.

Since, therefore, a clavicle in any degree of development is not essential to a climbing quadruped, we must seek for some other relation and use of that remarkably strong, and perfect bone, as it exists in the Megathere, Megalonyx, and Scelidothere. The absence of 'dentes primores' or of anterior or incisive teeth in these quadrupeds at once sets aside any idea of its connection with an action of the fore extremities, very common in the mammals which possess clavicles, viz., that of carrying the food to the mouth, and holding it there to be gnawed by the teeth. Flying is of course out of the question, although our surprise would hardly be less at seeing a beast as bulky as an elephant climbing a tree, than it would be to witness it moving through the air. If now we restrict our comparison to the relations of the clavicle in that order of Mammalia to which the extinct species in question belonged, we shall see that it is most constant, strongest, and most complete in those species which make most use of their strong and long claws in displacing the earth, as the Armadilloes and Orycteropus: and, as the clavicle is incomplete in one climbing Edental, we are naturally led to conclude that its perfect development in an extinct species must have been associated with uses and relations analogous to those with which it coexists in other genera of the same order. Thus it will be seen, that, in rejecting the conclusion drawn by M. Lund from the presence of a clavicle, I concur in the opinion expressed by Dr Buckland[53] that the Megatherium – and with it the Megalonyx and Scelidotherium – had the shoulder-joint strengthened by the clavicle, in reference to the office of the fore-arm, as an instrument to be employed in digging roots out of the ground. Not, however, that these gigantic quadrupeds fed on roots, but rather, as the structure of the teeth would show, on the foliage of the trees uprooted by the agency of this powerful mechanism of the fore-legs, and of the otherwise unintelligible colossal strength of the haunches, hind-legs, and tail.

The humerus presents a large convex oval head, on each side of which is a tuberosity for the implantation of the supra- and sub-scapular muscles: these tuberosities do not rise above the articular convexity, so as to restrict the movements of the shoulder-joint, as in the Horse and Ruminants, but exhibit a structure and disposition conformable to those which characterize the proximal extremity of the humerus in other mammalia which enjoy rotatory movements of the upper or fore-limb. The tuberosities are, however, relatively more developed, and give greater breadth to the proximal end of the humerus in the Scelidothere than in the Megathere. The distal end of the humerus, although mutilated, clearly indicates that it had the same characteristic breadth of the external and internal / condyles, as in the Megatherium. In fig. 1, Pl. XXV, which gives a front view of the left humerus, the broad internal condyle, with its extremity broken off, is seen projecting to the left hand; both in this figure and in fig. 2 in which the internal side of the humerus is turned towards the observer, the wide groove, with its two osseous boundaries, is shewn, which plainly indicates that the left condyle was perforated for the direct passage of the artery or median nerve, or of both, to the fore-arm. The groove for the musculo-spiral nerve on the outer side of the humerus is over-arched at its upper part by a strong obtuse process; which is comparatively less developed in the Megatherium. The trochlear or inferior articular surface of the humerus presents, as in the Megatherium, two well-marked convexities, with an intervening concavity: this indication of the rotatory power of the fore-leg is confirmed by the form of the head of the radius.

In Pl. XXV, fig. 4, a view is given of this articular surface: it presents

[53] *Bridgewater Treatise*, p. 152.

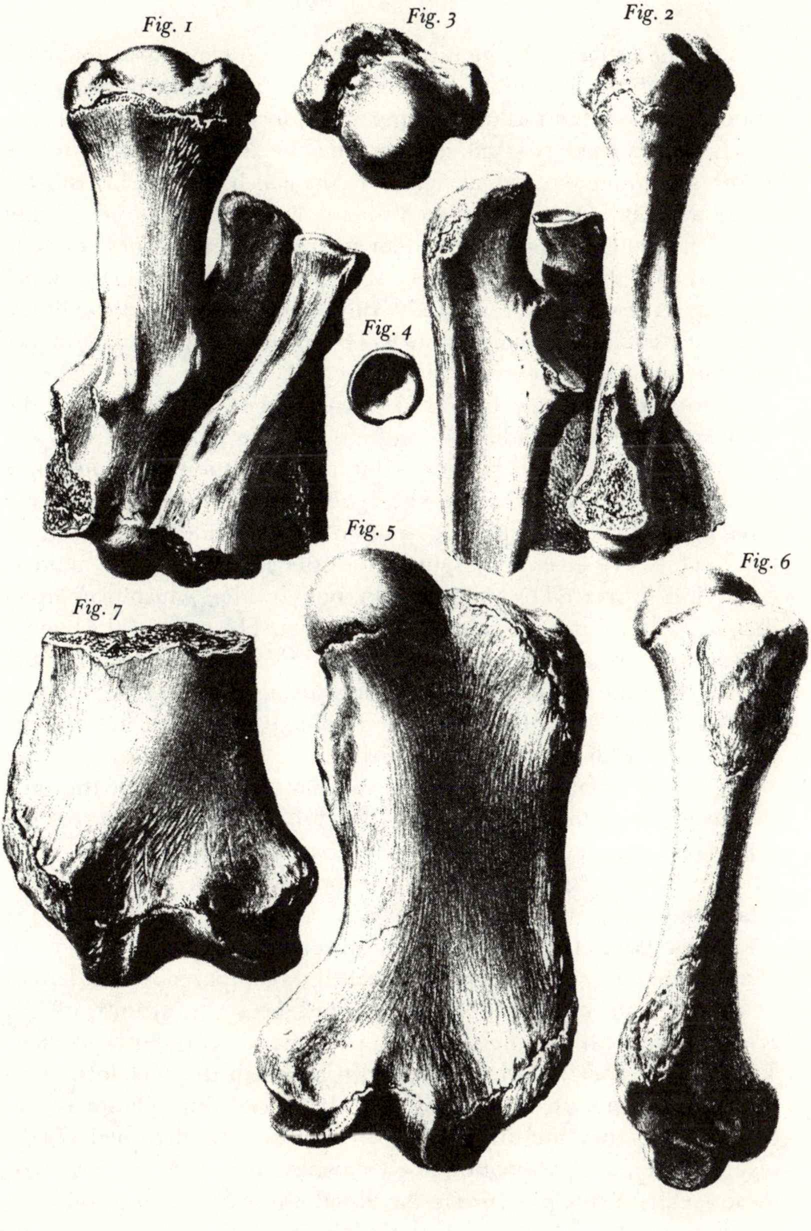

PLATE XXV

Scelidotherium

the form of a subcircular gentle concavity, which plays upon the outer convexity of the humeral articular surface: immediately below the upper concavity the radius presents a lateral smooth convex surface, which rotates upon a small concavity on the ulna, analogous to the 'lesser semilunar', in human anatomy, in which the mechanism for rotation, so far as the upper joint of the radius is concerned, is not more elaborately wrought out than in the present extinct edentate quadruped. The radius expands as it proceeds to the elbow-joint, where it attains a breadth indicative of the great power and size of the unguiculate paw, of which it may be called the stem, and to the movements of which it served as the pivot.

All the bones of the fore-limb just described – the scapula, the humerus, and the radius – indicate by the bold features and projections of the muscular ridges and tubercles the prodigious force which was concentrated upon the actions of the fore-paw, and the ulna, in its broad and high olecranon (of which a side-view is given in fig. 2, Pl. XXV) gives corresponding evidence. The great semilunar concavity is traversed by a sub-median smooth ridge, which plays upon the interspace of the two humeral convexities. The body of the bone is subcompressed, straight, and diminishes in size as it approaches the carpal joint: the immediate articulating surfaces are wanting in both the radius and ulna, the epiphysial distal extremities having become detached from their respective diaphyses.

Of the terminal segment of the locomotive extremities, the only evidence among the remains of the skeleton of the Scelidothere is the ungueal phalanx figured at Pl. XXVII, 3, 4, and 5; but as it is uncertain whether it belong to the fore or hind-foot, it will be described after the other bones of the extremities have been noticed.

Of these bones the femur is the most remarkable, both for its great proportional size, and its extreme breadth, as compared with its length or thickness: / but in all these circumstances the affinity of the Scelidothere with the Megathere is prominently brought into view. There is no other known quadruped with which the Scelidothere so closely corresponds in this respect. In proceeding, however, to compare together the thigh-bones of these two extinct quadrupeds, several differences present themselves, which are worthy of notice: of these the first is the presence in the Scelidothere of a depression for a 'ligamentum teres' on the back part of the head of the femur, near its junction with the neck of the bone: this is shewn in the posterior view of the femur given in Pl. XX. The head itself forms a pretty regular hemisphere: the great trochanter does not rise so high as in the Megatherium, but, relatively, it emulates it in breadth: the small trochanter is proportionally more developed: the external contour of the shaft of the femur is straighter in the Scelidothere than in the Megathere, and the shaft itself is less bowed forwards at that part. The articular condyles occupy a relatively smaller space upon the distal extremity of the femur in the Scelidothere, and they differ more strikingly from those of the Megathere, in being continued one into the other: the rotular surface, for example, which is shewn in fig. 5, Pl. XXV, is formed by both condyles, while in the Megatherium it is a continuation exclusively of the external articular surface.

The patella, which works upon the above-mentioned surface, is a thick strong ovate bone, with the smaller end downwards: rough and convex externally, smooth on the internal surface, which is concave in the vertical and convex in the transverse directions.

Of the bones of the leg only the proximal end of the tibia is preserved; but this is valuable, as shewing another well-marked difference between the Scelidothere and Megathere; for whereas in the latter the fibula is anchylosed with the tibia, this bone, in the Scelidothere, presents a smooth flat oval articular surface, which is shewn in fig. 2, Pl. XXVI, below the outer part of the head of the bone; from the size and appearance of which, I infer, that the fibula would not have become confluent with the tibia, even in the mature and full-grown animal.

The relative length of the fore and hind extremities cannot be precisely determined from the present imperfect skeleton of the Scelidothere; but there is good evidence for believing, that the fore extremity was the shortest. The humerus is shorter than the femur by one-ninth part of the latter bone; and the radius, which wants only the distal epiphysis, must have been shorter than the humerus. Now the relative development of the fore and hind legs is one of the points to be taken into consideration in an attempt to determine the habits and nature of an extinct mammal.

In climbing animals the prehensile power is more essential to the hinder than to the fore parts or extremities. In the leech the principal sucker is in the tail; / and higher organized climbers, in like manner, depend on their posterior claspers in descending trees, and hold on by means of them whilst selecting the place for the next application of those at the fore part of the body, whether their place be supplied by the beak, as in the Maccaws, or the fore-feet or hands in the Mammalia.

Fig. 1

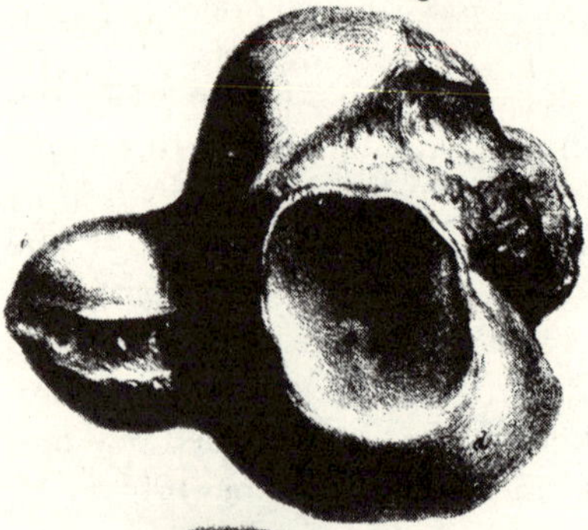

Fig. 2

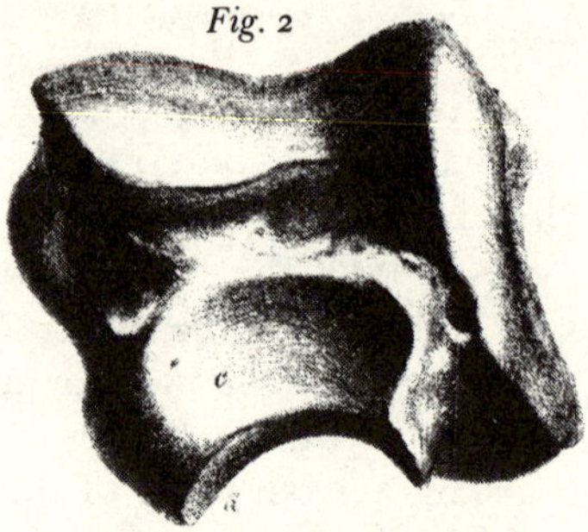

Fig. 3

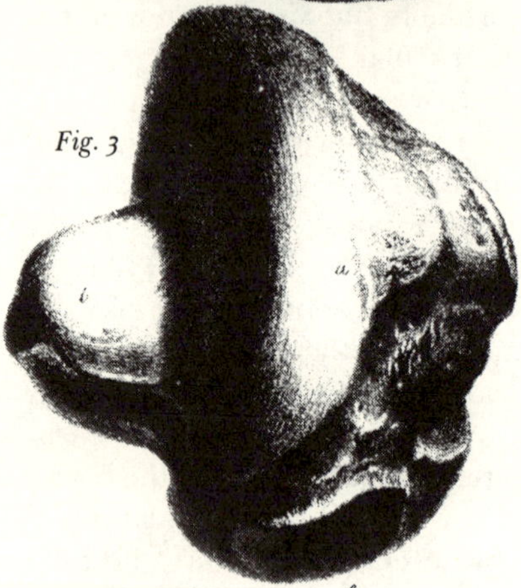

Fig. 4

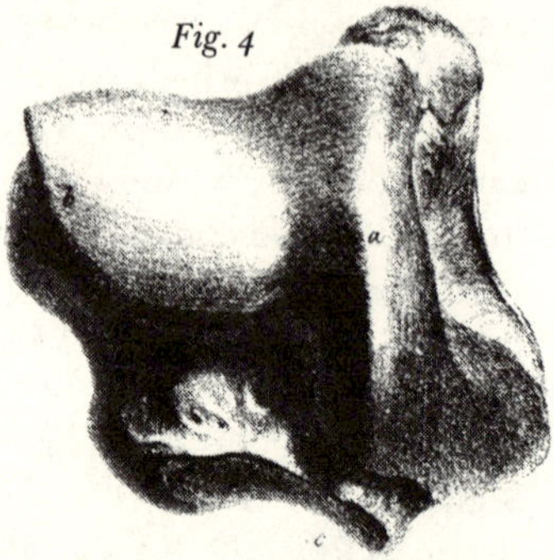

Fig. 5

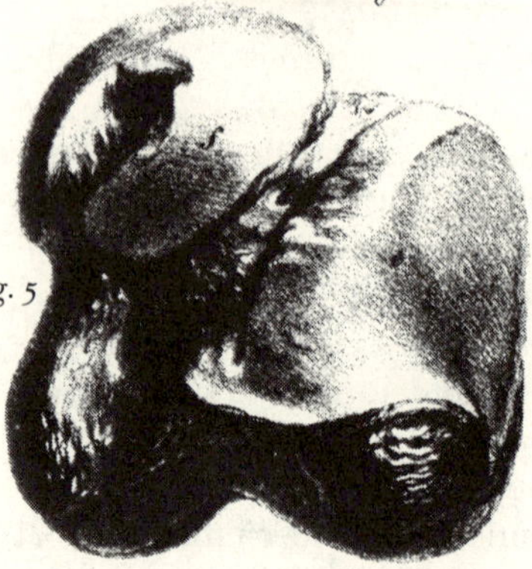

Fig. 6

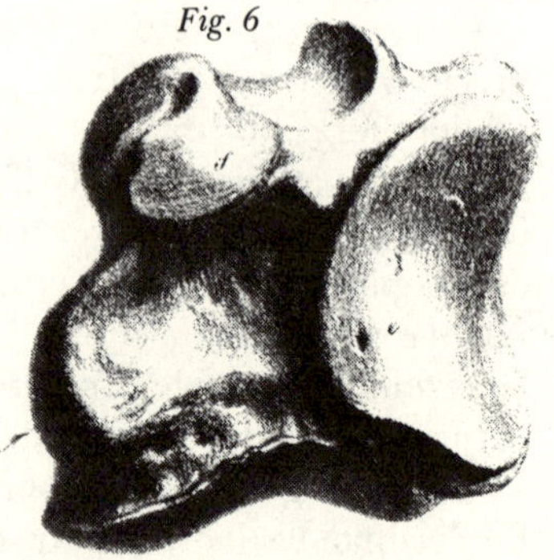

PLATE XXVI

Left astragalus

Figs 1, 3, 5. Megatherium *Figs 2, 4, 6. Scelidotherium*
Figs 1, 3 and 5 are drawn ½ the size of Figs 2, 4 and 6

But, although we perceive the hinder limbs to be the last to lose the advantageous structure of the hand in the Quadrumanous species, and notwithstanding that the tail is for this purpose sometimes specially organized to serve as a prehensile instrument, yet we find that the power of grasping the branches of trees by either legs or tail is never maintained at the expense of undue bulk and weight of those organs. On the contrary, as the fore-limbs are the main instruments in the active exertions of climbing, so they are the strongest as well as the longest in all the best climbers, and the weight of the body which they have to drag along is diminished by dwarfish proportions of the hinder limbs, as in the Orangs and the Sloths.

Can those huge quadrupeds have been destined to climb that had the pelvis and hinder extremities more ponderous and bulky in proportion to the fore-parts of the body than in any other known existing or extinct vertebrate animals?

M. Lund argues for the scansorial character of the Megalonyx, because its anterior extremities are longer than the posterior ones; but if they somewhat exceed the hind-legs in length, how vastly inferior are they in respect of their breadth and thickness. The prehensile faculty of the hinder limbs of the best climbers, as the Sloths, Orangs, and Chameleons is by no means dependent on the superior mass of muscle and bone which enters into their conformation, but is associated with the very reverse conditions.

It is impossible to survey the discrepancy of size between the femur and the humerus of the Scelidothere, as exhibited in Pl. XX, without a conviction that it relates to other habits than those of climbing trees. The expanse of the sacrum, the evidence of the muscular masses employed in working the hind legs and tail, which is afforded by the capacity of the cavity lodging the part of the spinal marrow from which the nerves of those muscles were derived, both indicate the actions of the hind-legs and tail to have been more powerful and energetic than would be required for mere prehension: and the association of hinder extremities so remarkable for their bulk, with a long and powerful tail, forbids my yielding assent to the speculation set forth by M. Lund, as to the prehensile character of the tail of the Megalonyx.

Astragalus. In the examination of this characteristic bone I have kept in view the question of the habits of the Megatherioid quadrupeds in general, and the especial affinities of the Scelidotherium, in illustration of which I shall notice at / the same time the peculiarities of the astragalus of the Sloth, Megatherium and Armadillo.[54]

The upper articular surface of the astragalus of the Scelidotherium (Pl. XXVII, fig. 4), presents, in its transverse contour, two convex pulleys, *a* and *b*, and an intermediate concavity, forming one continuous articular surface. The external or fibular trochlea (*a*) is strictly speaking convex only at its posterior part, the upper surface gradually narrowing to a ridge, as it advances forwards, from which, the inner and outer parts slope away at an angle of 35°.

The tibial[55] convexity (*b*) is more regular and less elevated, it has only half the antero-posterior extent of the outer pulley; its marginal contour forms an obtuse angle at the inner side.

In the Megatherium the upper articular surface of the astragalus is also divided into two trochleae, of which the one on the fibular side (fig. 3, *a*), is of much greater relative size and extent than the tibial one (*b*), and is raised nearly four inches above the level of the latter, although in the oblique position in which the bone is naturally placed in the skeleton, the highest part of each convexity is on the same level. The fibular trochlea differs also from that in the Scelidothere in being regularly convex in the transverse as well as the antero-posterior direction. The tibial convexity resembles that in the Scelidothere, save in its smaller relative size; its internal margin likewise forms an angular projection below the internal malleolus.

The upper surface of the astragalus of the Mylodon, or Megalonyx(?) (Pl. XXVIII, fig. 5),[56] differs from that in the Megatherium in having a narrower fibular trochlear ridge.

The astragalus of the Ai (*Bradypus tridactylus*) differs widely from that of either the Megathere, Mylodon(?) or Scelidothere in having a conical cavity on the upper surface, in place of the fibular convexity, in which concavity the distal end of the fibula rotates like a pivot. This

[54] *Dasypus 6-cinctus*, L., is the species of which I have the astragalus separate, so as to be able to follow out the comparison.

[55] In distinguishing these trochleae as fibular and tibial, it is to be understood that the terms relate only to aspects corresponding to the position of those bones, and not that the fibula is articulated to the whole of the trochlea so called: it probably rested only upon the outer facet in the Scelidothere.

[56] This astragalus was found at Santa Fé, in Entre Rios, associated with the remains of the Mastodon and Toxodon; but from its size and form I entertain little doubt that it belonged to a Megatherioid quadruped as large as the Mylodon or Megalonyx. The brief allusion to the astragalus of the Megalonyx in M. Lund's *Memoir* does not afford the means of determining with certainty this point.

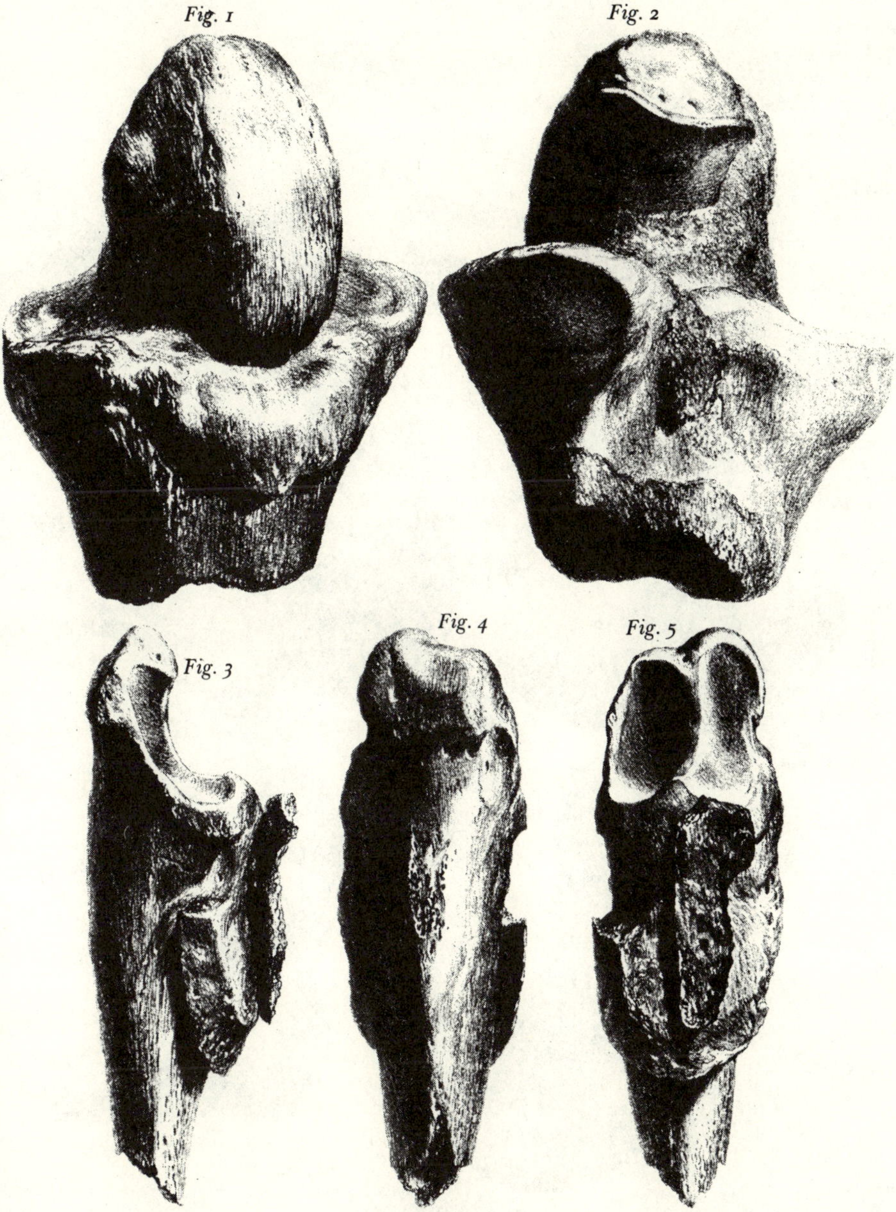

PLATE XXVII

Scelidotherium

Figs 1, 2 are drawn ⅔rds the size of Figs 3–5

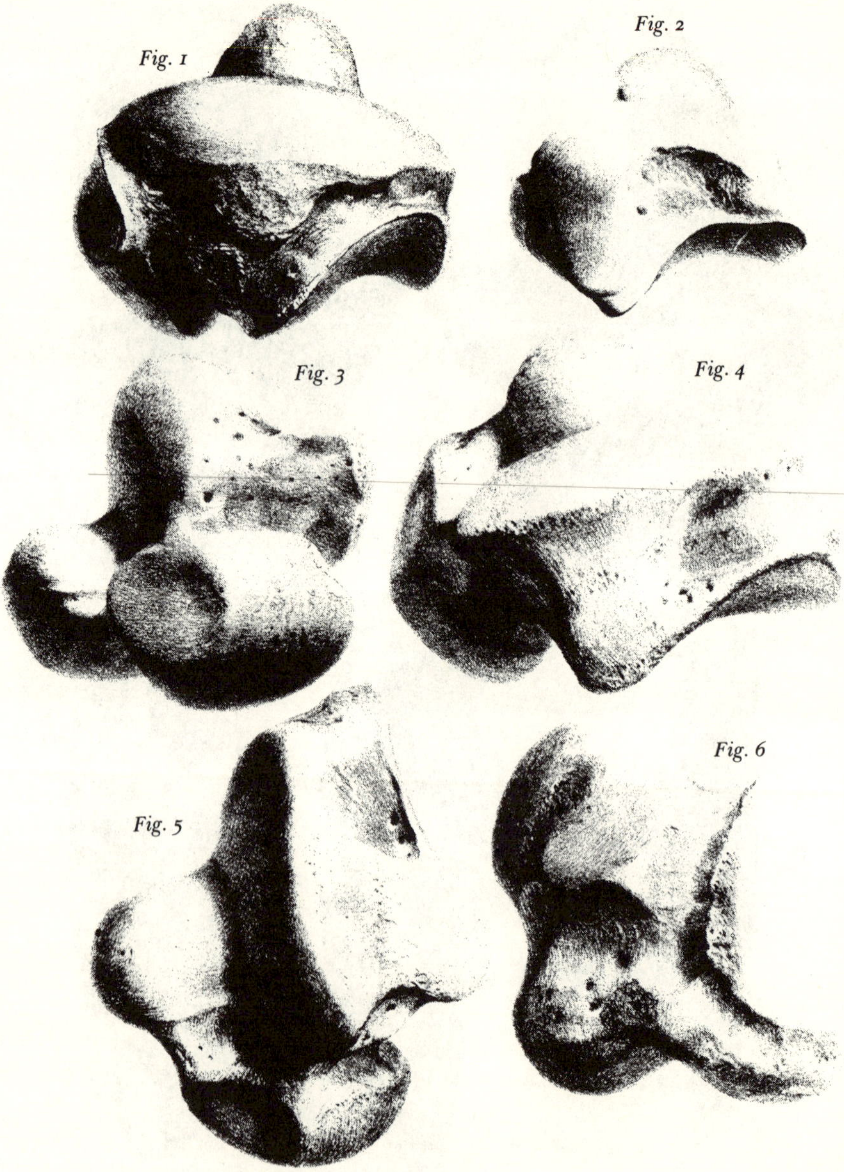

PLATE XXVIII

Left astragalus

Fig. 1. Megatherium *Fig. 2. Scelidotherium* *Figs 3–6. Mylodon?*
Fig. 1 is drawn ½ the size of Figs 2–6

mechanism is closely related to the scansorial uses of the inwardly inflected foot of the Sloth.

If the astragalus of an Armadillo[57] were placed side by side with that of the / Megathere, it would be very difficult to determine the analogous parts, especially of the upper surface, unless guided by the intermediate structure presented by the Scelidothere. The upper surface of this bone, in the Armadillo, is, however, divided into two transversely convex trochleae, separated by a much wider transversely concave surface. The fibular trochlea resembles that of the Scelidothere in having its upper and outer facets sloping away at an acute angle, but without meeting at a ridge anteriorly; this surface is not more raised above the tibial trochlea than in the Scelidothere.

The inner trochlea differs from that of the Scelidothere in having a greater relative antero-posterior extent, and in forming, in place of an uniform convex surface, a trochlea similar in structure to that on the outer side. The extent of rough surface on the upper part of the astragalus intervening between the articular surface for the bones of the leg, and that for the scaphoides is extremely small in the Megathere and Mylodon(?); it is relatively greater in the Scelidothere; it is still more extensive in the Armadillo; but is the longest in the Sloth. The anterior extremity of the astragalus which is entirely occupied by the scaphoid articular surface is very peculiar in the Scelidothere (Pl. XXVII, fig. 2): it presents one convex and two concave facets, which, however, form part of one continuous articular surface: the convex facet forms the internal part of the surface, and presents a rhomboidal form with the long axis vertical. The concave facets (*c* and *d*) are extended transversely and placed one above the other; they are slightly concave in the transverse, and nearly flat in the vertical directions.

In the Megatherium (fig. 1) the scaphoid surface of the astragalus is divided only into one concave and one convex portion, both continuous with each other: the concave facet (*c*) corresponds with the upper concavity in the Scelidothere, but is a pretty uniform subcircular depression, fourteen lines in depth: the convex facet, *d*, is continued across the whole breadth of the under part of the scaphoid surface and corresponds with both the inner convex, and lower concave surfaces of the scaphoid articulation in the Scelidothere.

[57] See the figures of this bone, given by Cuvier in Pl. x and xi, *Ossemens Fossiles*, vol. v, part i.

In the Mylodon(?) (Pl. XXVIII, fig. 3), the articular facet, corresponding with that marked (*c*) in the astragali of the Megathere and Scelidothere, is simply flattened, instead of being concave; the rest of the scaphoid surface corresponds with that in the Megatherium.

In the Armadillo the scaphoid articular surface is undivided and wholly convex: in this part of the astragalus, therefore, we find the Scelidothere deviating from the Armadillo further than does the Megathere; while the Mylodon or Megalonyx(?) most resembles the Armadillo in the configuration of this part of the astragalus.

If we compare the outer surfaces of the astragalus in these quadrupeds, / we shall find, however, that the Scelidothere and Armadillo closely agree: the outer facet of the fibular trochleae, above described, is continued in the Scelidothere (Pl. XXVIII, fig. 2), upon the fibular side of the astragalus reaching nearly half-way down the posterior part, and down nearly the whole of its anterior.

In the Armadillo, it extends over the whole of the anterior part of the outer side of the astragalus. In both animals the lower boundary of this articular surface describes a strong sigmoid curve.

In the Megatherium (Pl. XXVIII, fig. 1), the corresponding surface for the fibular malleolus on the outer side of the astragalus is formed by a comparatively very small semicircular flattened facet, which by its roughness indicates that the end of the fibula was attached to it by ligamentous substance, and that the synovial bag was not continued upon that surface as in the Scelidothere and Armadillo.

In the Mylodon(?) (Pl. XXVIII, fig. 4), even this rough facet is wanting and the fibular trochlea is bounded by the angle which divides the upper from the outer surface of the astragalus.

Turning now our attention to the under surface of the astragalus, we observe that it presents in the Scelidothere (Pl. XXVI, fig. 6), an irregular quadrate form, having the outer side occupied by an elongated sub-ovate articular facet, *e*, for the calcaneum, bounded externally by a sharp edge, with its long axis and its greatest concavity in the antero-posterior direction, and slightly convex from side to side: a second calcaneal articular surface (*f*) is situated at the inner and anterior angle; it is oblong and nearly flat; is continuous with the inferior concave facet of the scaphoid articulation, but is divided from the convex facet by a groove: the two calcaneal articulations are separated by a deep and rough depression, traversing the under surface of the astragalus diagonally, and increasing in breadth towards the posterior and internal angle. The inner side of the astragalus presents a convex protuberance.

The correspondence between the astragalus of the Scelidothere and Megathere is best seen at the under surface of the bone: in both the two calcaneal articulations are separated by the diagonal depression, and the internal and anterior surface is continuous with the scaphoid articulation. In the Megathere, however, in consequence of the absence of the inferior concavity which characterizes the Scelidothere, the anterior calcaneal facet (*f*) appears as a more direct backward continuation of the scaphoidal surface; but they are divided by a more marked angle than is represented in the figure (fig. 5, Pl. XXVI). The posterior and outer calcaneal surface in the Megathere (*e*) is broader in proportion to its length, continued further upwards upon the outward surface, is consequently more convex in the transverse direction, and is not bounded externally by so sharp and prominent a ridge as in the / Scelidothere. The protuberance from the inner surface of the astragalus is more compressed laterally in the Megathere than in the Scelidothere. The correspondence between the astragali of the Mylodon(?) (Pl. XXVIII, fig. 6) and Megathere in the conformation of the under surface is so close, that the few differences which exist will be sufficiently appreciated by an inspection of the figures.

In the Armadillo the astragalus, in consequence of the greater production of its anterior part, presents more of an angular than a quadrate figure; and the scaphoid articular surface, being proportionally carried forwards, is altogether separated from the anterior calcaneal surface. The posterior and inner calcaneal surface resembles that in the Scelidothere, but is less inclined upwards; and is continuous with the posterior part of the tibial articular surface.

Thus the astragalus in the structure of its two most important articulations, viz. that which receives the superincumbent weight from the leg, and that which transmits it to the heel, presents a closer correspondence in the Scelidothere with that of the Dasypus, than with that of the Megathere or Mylodon.

The ungucal phalanx of the Scelidothere before alluded to, is represented of the natural size in Pl. XXVII. The side-view, fig. 3 shows the position of the articular surface on the proximal end, sloping obliquely towards the under surface, and overtopped by an obtuse protuberance, calculated to impede any upward retraction of the claw: the present joint, in fact, illustrates in every particular the argument by which Cuvier established the true affinities of the allied extinct genus Megalonyx.[58]

The present phalanx, is, however, less compressed, and less incurved than those of the Megalonyx, which have been hitherto described; but it more resembles in these proportions one of the smaller, and presumed hinder, ungueal phalanges of the Megatherium. The upper and lateral parts of the bone are rounded, and it gradually tapers to the apex, which is broken off. The osseous sheath for the claw is developed only at the under part of the bone: it presents the form of a thick flat plate of bone, with the margin very regularly and obliquely bevelled off, and having a vertical process of bone attached lengthwise to the middle of its under surface. This process must have served for the insertion of a very powerful flexor tendon. The figures of this bone preclude the necessity of any further verbal description.

M. Lund lays most stress upon the argument founded on the inward inflection of the sole of the foot in the Megalonyx, and appeals with greatest confidence to this structure in support of his hypothesis of the scansorial habits of that extinct Edental.[59] /

[58] *Ossemens Fossiles*, vol. v, part i, p. 163.

[59] For the translation of the following passage, and of others alluded to in the present work, from the original *Danish Memoir* of M. Lund, loc. cit., I am much indebted to the Rev. W. Bilton, M.A., etc.: /

'Thus in every point of comparison we have instituted between the organization of burrowers and climbers; we have seen that the Megalonyx constantly differs from the former and resembles the latter; but the point to which I last alluded (the obliquity of foot), I consider to be quite decisive.

'There is one other point in its organization, which is not quite without weight in reference to our present inquiry – I mean its unusually powerful tail. Now, it is certainly true that many animals which are not climbers have a powerful tail, as e.g. Armadillos, while the others that climb well, have none, as Sloths and Apes. But when we find a remarkably powerful tail attached to an animal that according to all probability was a climber, we are led to infer that this organ must have served for that purpose: in other words, that the Megalonyx was furnished with a prehensile tail.

'How far the Megatherium is to be considered in the same light as the Megalonyx cannot be decided without an accurate and scientific examination of its skeleton at Madrid. Pander and D'Alton do not mention any distortion of the hind-foot, neither does their figure exhibit any. It is nevertheless quite possible that such may exist, but that it is disguised by the faulty manner in which the skeleton is put up. It strikes me as little probable that two animals which agree so well in the principal particulars of their organization should differ so much in one of the most important. The Megatherium has been proved by later discoveries to possess the same powerful tail as the Megalonyx, and as it corresponds also with the latter entirely in the conformation of its extremities, the same difficulties present themselves against the supposition of its having been a burrower. But if the Megatherium was really a climber, it must have had still more occasion (on account of its greater size), for that peculiar arrangement of the hind-feet which we have described in the Megalonyx.'

It is quite true that the Quadrumana derive advantage from this position of the foot in climbing trees, and that it is carried to excess in the Sloths, which can only apply the outer edge of the foot to the ground. But we may ask, was the inversion of the sole of the foot actually carried to such an extent in the *Megalonyx*? And, admitting its existence in an inferior degree, is it then conclusive as to the scansorial habits of that species?

M. Lund expressly states that it is produced by a different structure and arrangement of the tarsal bones, from that which exists in the Sloth, but he does not specify the nature of this difference.

If the astragalus, which I have referred with doubt to the *Megalonyx*, do not actually belong to that genus, it is evidently part of a very closely allied species. Now this astragalus, as we have seen, resembles most closely that of the Megatherium; and since we may infer that the calcaneum, scaphoides, and cuboides had a like correspondence, the inclination of the sole of the foot inwards must have been very slight, as I have determined from examination of the structure and co-adaptation of those bones in the incomplete skeleton of the Megatherium in the London College of Surgeons. Such an inclination of the foot may be conceived to have facilitated the bending of the long claws upon the sole, during the ordinary progressive movements of the animal, but it is quite insufficient to justify the conclusion, that it related to an application of the hind feet for the purposes of climbing.

It is not without interest again to call to mind the deviation of the structure / of the astragalus of the Scelidothere from the Megatherioid to the Dasypodoid type of structure. For if the Megatherioid type of structure had really been one suitable to the exigencies of climbing quadrupeds, it might have been expected to have exhibited the scansorial modifications more decidedly, as the species diminished in stature; but as regards the instructive bone of the hind-foot, the modifications of which we have just been considering, this is by no means the case.

DESCRIPTION OF A MUTILATED LOWER JAW OF THE

MEGALONYX JEFFERSONII

In the preceding section an astragalus was described, which was regarded as belonging possibly to the same Edentate species as the jaw figured and described, p. [69] 80, Pl. XVIII and XIX, under the name of *Mylodon Darwinii*; but the same correspondence – that of relative size – renders it equally possible that this astragalus may belong to the species of the *Megalonyx* to which the lower jaw now under consideration appertains. There could be no doubt, from its structure, that it was the astragalus of a gigantic species of the order *Bruta*, and of the *Megatherioid* family, and more nearly allied to the Megathere than is the Scelidothere, but sufficiently distinct from both.

The lower jaw, figured in Pl. XXIX, is the only fossil brought home by Mr Darwin that could be confidently referred to the genus *Megalonyx*; but the form of the tooth in place on the right side of the jaw fully justifies this determination. The jaw itself is deeply and firmly imbedded in the matrix, so that only the upper or alveolar border is visible. The coronoid and condyloid processes are broken away, and the texture of the remaining part of the jaw was too friable, and adhered too firmly to the surrounding matrix to admit of more of its form being ascertained than is figured.

There were four molars on each side of this jaw; the large oblique perforation near the fractured symphysis is the anterior extremity of the wide dental canal. The forms of the alveoli are best preserved in the right ramus: the first is the smallest, and seems to have contained a tooth, of which the transverse section must have been simply elliptical: the second tooth is likewise laterally compressed, but the transverse section is ovate, the great end being turned forwards: the third socket presents a corresponding form, but a larger size: the fourth socket is too much mutilated to allow of a correct opinion being formed as to the shape of the tooth which it once contained. The natural size of the tooth / *in situ*, and of the adjoining socket, is given in Pl. XXIX, fig. 2. The difference of form which the jaw of the Megalonyx presents, as compared with that of the Mylodon, especially in the greater recedence of the two horizontal rami from each other, will be appreciated by comparing Pl. XVIII with Pl. XXIX.

DESCRIPTION OF A FRAGMENT OF THE SKULL AND OF THE TEETH OF THE

MEGATHERIUM CUVIERI

Notwithstanding the full, accurate, and elaborate accounts of the

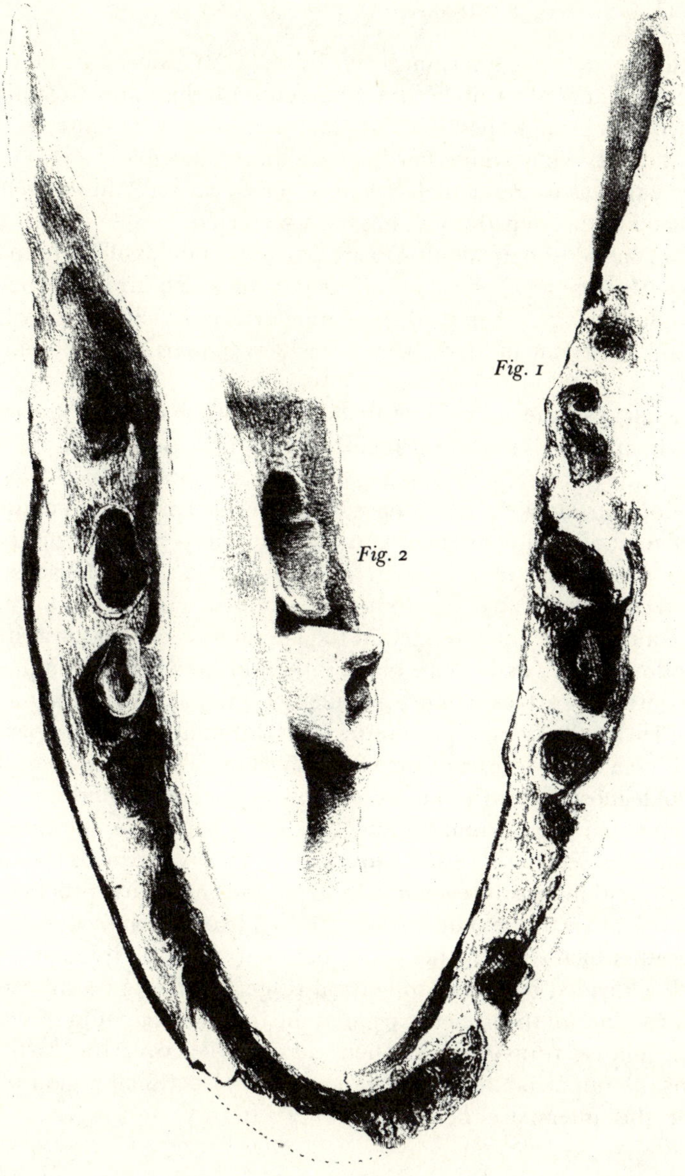

PLATE XXIX

Lower jaw of Megalonyx

Fig. 1 is drawn 2/3rds the size of Fig. 2

skeleton of the Megatherium given by Brû,[60], Cuvier,[61] Pander and D'Alton,[62] and Mr Clift,[63] the fragments of this most gigantic of quadrupeds brought home by Mr Darwin, possess much interest, and have added, what could hardly have been anticipated, important information as to the dental system, whereby an error in the generic character of the Megatherium has been corrected.

The fragments here alluded to are portions of the skull of three full-grown Megatheres:, the most perfect part of which affords a view of the posterior, and of part of the basal surface, which regions of the cranium have not hitherto been elsewhere figured or described (Pl. XXX).

The plane of the occipital foramen forms with that of the base of the skull an angle of 140°, the plane of the posterior surface of the skull forms with the basal plane an angle of 68°. The occipital condyles are therefore terminal, or form the most posterior parts of the cranium. The extent of their convex curvature in the antero-posterior direction, which equals that of a semicircle, indicates that the Megatherium possessed considerable freedom and extent of motion of the head. The condyles are not extended in the lateral direction so far as in the Toxodon; their axis is more oblique than in the Glossotherium, and their internal surface is more parallel with the axis of the skull, the foramen magnum not presenting that infundibuliform expansion which is so characteristic of the Glossotherium. The occipital condyles resemble most in form and position those of the Scelidotherium; but in the angle of the occipital plane the Megatherium is intermediate between the Scelidothere and Glossothere. The ex-occipitals terminate laterally and inferiorly, each in a short, but strong obtuse process. The posterior plane of the skull is traversed by a strong arched intermuscular crest, which / forms the upper boundary of a pretty deep fossa, which is divided by a median vertical ridge, extending downwards to within an inch of the upper margin of the foramen magnum. A second strong obtuse transversely arched ridge curves over the first, and forms the upper boundary of the posterior or occipital region of the skull: the interspace between the two transverse ridges is very

[60] Descripcion del Esqueleto de un quadrupedo muy corpulento y raro, que se conserva en el Real Gabinete de Historia Natural de Madrid. *Folio*, Madrid, 1796.

[61] *Ossemens Fossiles*, tom. v, pt i, p. 179.

[62] Das Riesen Faulthier, Bradypus giganteus, von Dr Chr. Pander und Dr E. D'Alton. *Folio*, Bonn, 1821.

[63] *Transactions of the Geological Society*, 1835, p. 438.

Fig. 1

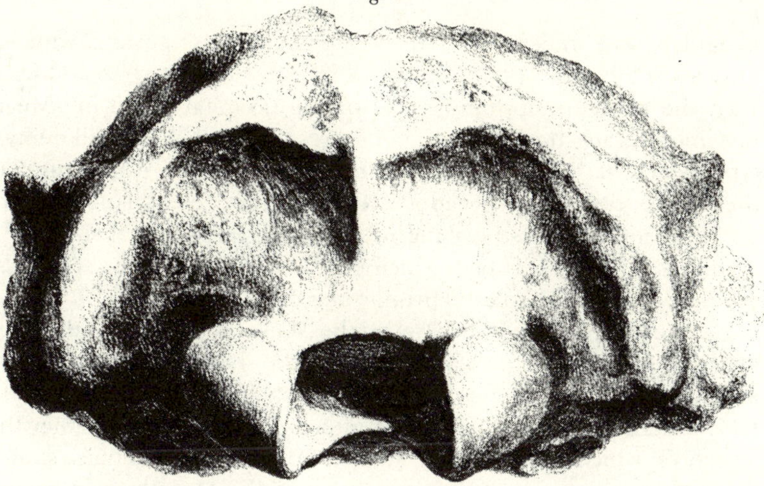

Fig. 2

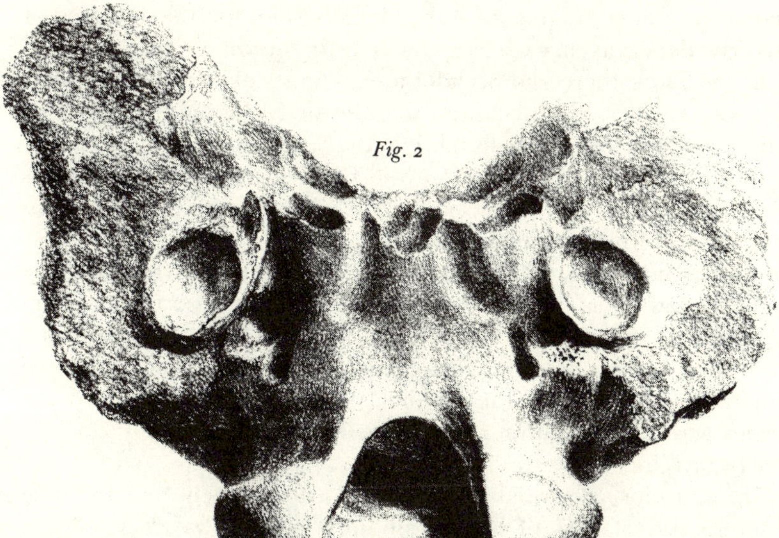

PLATE XXX

Megatherium

irregular, and indicates the firm implantation of powerful nuchal muscles or ligaments (Pl. XXX, fig. 1).

In the configuration and angle of the occipital plane the Megaherium indicates the same general correspondence with the Edentate type, which has been pointed out in the descriptions of the crania of the Glossothere and Scelidothere: and the resemblance to the Scelidothere is not less striking in the small proportional size of the cranium in this quadruped, which surpasses the rest of its class in so great a degree in the colossal proportions of its hinder parts.

Having detected in the base of the skull of the Scelidothere an articular semicircular pit for the head of the styloglossal bone, similar to, but relatively smaller than, that remarkable one in the skull of the Glossothere, it became a matter of interest to determine whether this structure, which does not exist in any of the existing Edentals, should likewise be present in the gigantic type of the Megatherioid family. The result of a careful removal of the matrix from the basal region of one of the cranial fragments of the Megatherium was the detection of this articular cavity, in each temporal bone in the same relative position as in the Glossothere and Scelidothere. The styloid articular cavity is relatively smaller, and shallower, than in the Glossothere, its proportions being much the same as those of the Scelidothere. The cranial or posterior extremity of the stylo-hyoid bone in the Scelidotherium is bent upwards at an obtuse angle (Pl. XXI), and terminates in an articular ball which rotates in this cavity. The size of this bone, and its mode of articulation, indicates great power and muscularity of tongue in the Megatherioids, and calls to mind the importance of that organ in the Giraffe, which subsists on the same kind of food as that which I have supposed to have supported the Megatherioids, although the general organization of these animals and the mode in which the foliage was brought within reach of the tongue are as opposite as can well be imagined.

The anterior condyloid foramen presents scarcely one half the absolute size of that of the Glossothere, whence we may infer a correspondingly inferior development of the tongue in the Megathere. The fractured parietes of the cranial cavity of the Megatherium every where exhibit evidences of the great extent of the air-cells or sinuses continued from the nasal cavity: on the basilar aspect of the cranium they extend as far back as the jugular foramina: the whole of the basisphenoid being thus excavated, and permeable / to air, derived from the sphenoid sinuses (Pl. XXX, fig. 2). The vertical diameter of the cranial cavity is four inches, eight lines; its transverse diameter, which is greatest in the posterior third part of the cavity, corresponding with the posterior part of the cerebrum is six inches: from the indications afforded by the remains of the cranial cavity in Mr Darwin's specimens, I conclude that the brain of the Megatherium was more depressed, and upon the whole, smaller by nearly one-half than that of the Elephant; but with the cerebellum relatively larger, and situated more posteriorly with relation to the cerebral hemispheres: whence it may be concluded that the Megatherium was a creature of less intelligence, and with the command of fewer resources, or a less varied instinct than the Elephant.

It has been usual to characterize the Megatherium, in conformity with the concurrent descriptions of Bru, Cuvier, and D'Alton, by the dental formula of *molares* $\frac{4}{4}\frac{4}{4}$, i.e. by the presence of four grinding teeth on each side of the upper, as of the lower jaw. It was the agreement of the excellent authorities above cited in this statement, which induced Mr Clift and myself to regard a single detached tooth, which formed part of the valuable collection of remains of the Megatherium deposited in the Hunterian Museum by Sir Woodbine Parish, as being, from its comparatively small size, the tooth of either a younger individual or of a smaller species of Megatherium. Upon clearing away the matrix from the palatal and alveolar surface of one of the cranial fragments of the Megatherium in Mr Darwin's collection, I was gratified by the detection of the crown of a fifth molar, corresponding in size and form with the detached tooth, above alluded to: its small size, and its position have doubtless occasioned its being over-looked in the cranium of the great skeleton at Madrid.

The anterior molar of the upper jaw presents a nearly semicircular transverse section, with the angles rounded off; the three succeeding teeth are four-sided, with the transverse somewhat exceeding the antero-posterior diameter: they are rather longer and larger than the first: the last molar is likewise four-sided, but presents a sudden diminution of diameter, and is relatively broader. The following are the respective dimensions of the upper maxillary teeth.

	FIRST MOLAR		SECOND MOLAR		THIRD MOLAR		FOURTH MOLAR		FIFTH MOLAR	
	Inches	*Lines*	*Inches*	*Lines*	*Inches*	*Lines*	*Inches*	*Lines*	*Inches*	*Lines*
Length	8	6	9	4	9	4	8	7	5	2
Transverse diameter	1	9	2	4	2	3	2	0	1	4
Antero-posterior diameter	1	5	2	0	2	0	1	11	0	10

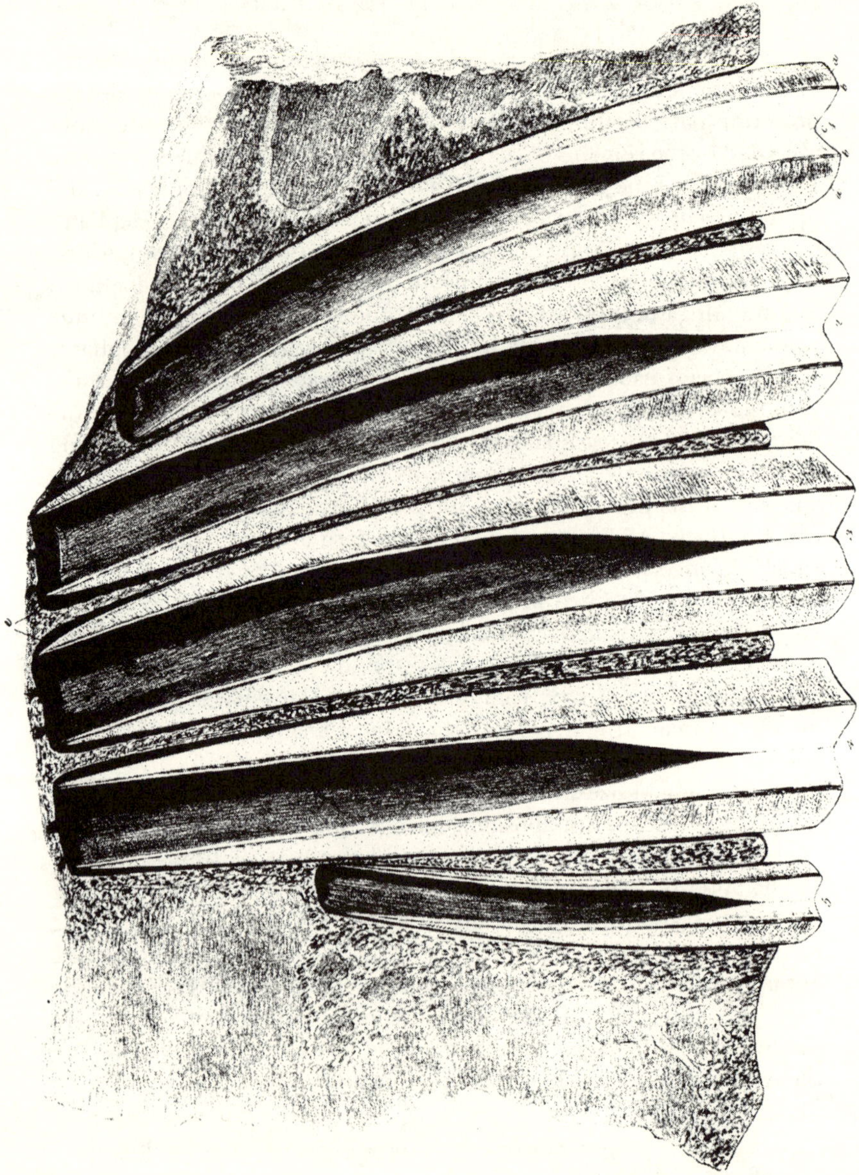

PLATE XXXI

Section of the upper jaw and teeth of Megatherium

Besides the differences in size, the upper molars vary as to their curvature: this difference is exhibited in the vertical section of these teeth figured in Pl XXXI. The convexity of the curve of the first, second and third molars is directed / forwards; the fourth is straight, its anterior surface only describing a slight convexity in the vertical direction; the fifth tooth is curved, but in a contrary direction to the others; and the bases of the five molars thus present a general convergence towards a point a little way behind the middle of the series.

The next peculiarity to be noticed in these remarkable teeth is the great length of the pulp-cavity (*d*), the apex of which is parallel with the alveolar margin of the jaw; a transverse fissure is continued from this apex to the middle concavity of the working surface of the tooth, which is thus divided into two parts. Each of these parts consists of three distinct substances – a central part analogous to the body or bone of the tooth or 'dentine', a peripheral and nearly equally thick layer of *caementum*, and an intermediate thinner stratum of a denser substance, which is described in Mr Clift's memoir on the Megatherium as 'enamel', and to which substance in the compound teeth of the Elephant, it is analogous both in its relative situation, and relative density to the other constituents.

Microscopic examinations of thin and transparent slices of the tooth of the Megatherium prove, however, that the dense layer separating the internal substance from the caementum is not enamel, but presents the same structure as the hard 'dentine' or ivory of the generality of Mammalian teeth; and corresponds with the thin cylinder of hard 'dentine' in the tooth of the Sloth. No species of the Order *Bruta* has true enamel entering into the composition of its teeth; but the modifications of structure which the teeth present in the different genera of this order are considerable, and their complexity is not less than that of the enamelled teeth of the Herbivorous Pachyderms and Ruminantia, in consequence of the introduction of a dental substance into their composition corresponding in structure with that of the teeth of the *Myliobates*, *Psammodus*, and other cartilaginous fishes.

The microscopic investigation of the structure of the teeth of the Megatherium was undertaken chiefly with the view of comparing this structure with that of the teeth of the Sloth and Armadillo, and of thus obtaining an insight into the food, and an additional test of the real nature of the disputed affinities of the Megatherium. The central part of the tooth (*c*. Pl. XXXI) consists of a coarse ivory, like the corresponding part of the tooth of the Sloth. It is traversed throughout by medullary canals 1/1500th of an inch in diameter, which are continued from the pulp-cavity, and proceed, at an angle of 50°, to the plane of the dense ivory, parallel to each other, with a slightly undulating course, having regular interspaces, equal to one and a half diameters of their own area, and generally anastomosing in pairs by a loop of which the convexity is turned towards the origin of the tubes of the fine dentine, as if each pair so joined consisted of a continuous reflected canal (*c.* fig. 1, Pl. XXXII). The loops are generally / formed close to the fine dentine. In a few situations I have observed one of the medullary canals continued across the fine dentine, and anastomosing with the corresponding canals of the caementum. The interspaces of the medullary canals of the coarse dentine are principally occupied by calcigerous tubes which have an irregular course, anastomose reticularly, and terminate in very fine cells. The more regular and parallel calcigerous tubes, which constitute the thin layer of hard dentine, are given off from the convexity of the terminal loops of the medullary canals. The course of these tubes (*b.* fig. 1, Pl. XXXII) is rather more transversely to the axis of the tooth than the medullary canals from which they are continued. They run parallel to each other, but with minute undulations throughout their course, in which they are separated by interspaces equal to one and a half their own diameter. As they approach the caementum they divide and sub-divide, and grow more wavy and irregular: their terminal branches take on a bent direction, and form anastomoses, dilate into small cells, and many are seen to become continuous with the radiating fibres or tubes of the cells or corpuscles of the contiguous caementum. This substance enters largely into the constitution of the compound tooth of the Megatherium: it is characterized, like the caementum of the Elephant's grinder, by the presence of numerous radiated cells, or purkingian corpuscles, scattered throughout its substance, but may be distinguished by wide medullary canals which traverse it in a direction parallel with each other, and forming a slight angle with the transverse axis of the tooth. These canals are wider than those of the central coarse dentine, their diameter being 1/1200th of an inch; they are separated by interspaces equal to from four to six of their own diameters, divide a few times dichotomously in their course, and finally anastomose in loops, the convexity of which is directed towards, and in most cases is in close contiguity with, the layer of dense dentine.

Fine calcigerous tubes are every where given off at right angles from

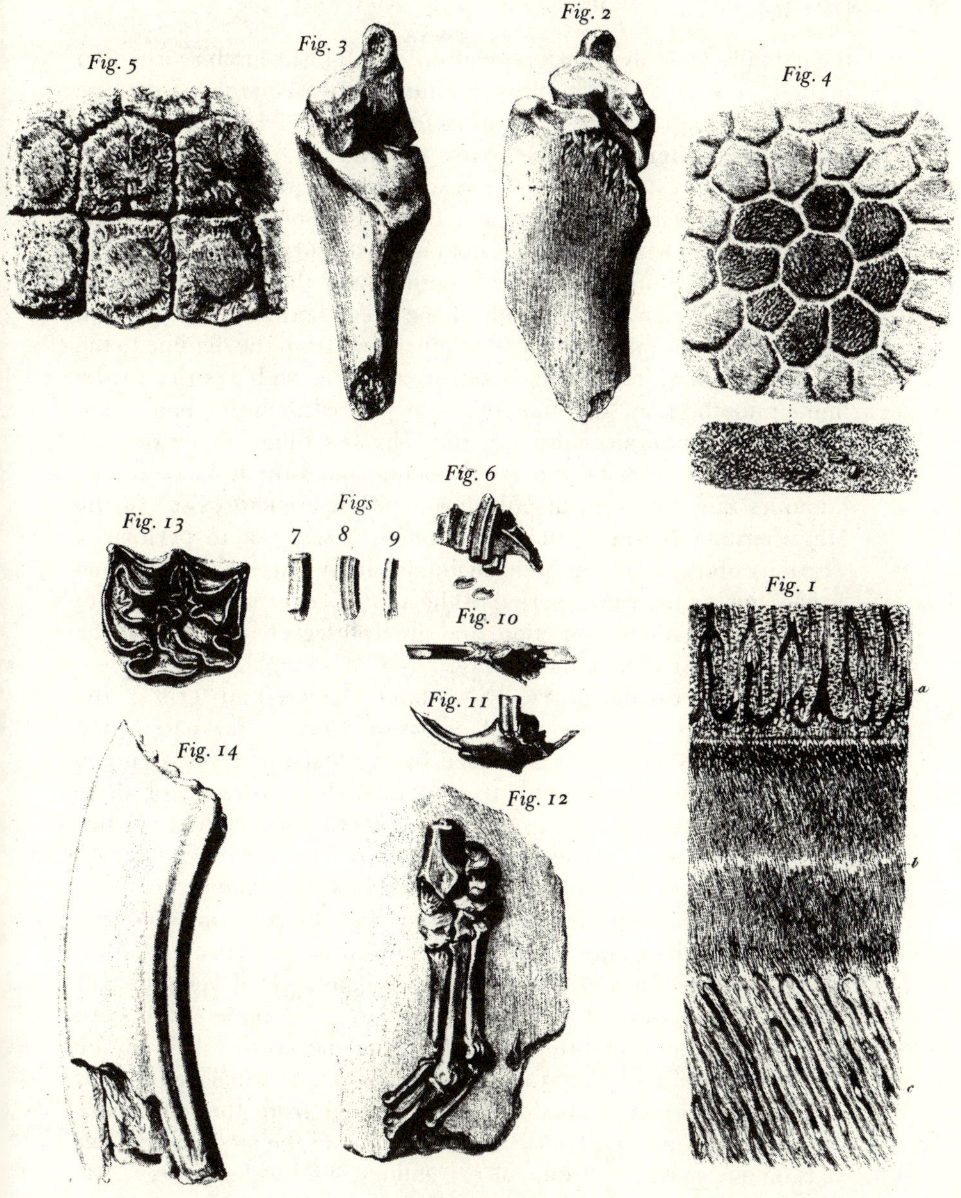

PLATE XXXII

Fig. 1. Megatherium *Figs 2–5. Hoplophorus* *Figs 6–12. Ctenomys*
Figs 13, 14. Equus

the medullary canals of the caementum, which form a rich reticulation in their interspaces, and a direct continuation between the loops of the medullary canals and the calcigerous tubes of the dense dentine. The caementum differs from the coarse dentine in the larger size and wider interspaces of its medullary canals, and by the presence of the bone-corpuscles in their interspaces; but they are brought into organic communication with each other, not only by means of the tubes of the dense dentine, but by occasional continuity of the medullary canals across that substance. The tooth of the Megatherium thus offers an unequivocal example of a course of nutriment from the dentine to the caementum, and reciprocally. Retzius observes with respect to the human tooth, that 'the fine tubes of the caementum enter into immediate communications with the cells and tubes of the dentine (zahnknochen), so that this part can obtain from without the requisite / humours after the central pulp has almost ceased to exist.' In the Megatherium, however, those anastomoses have not to perform a vicarious office, since the pulp maintains its full size and functional activity during the whole period of the animal's existence. It relates to the higher organized condition, and greater degree of vitality of the entire grinder in that extinct species.

The conical cavities (*d*. Pl. XXXI) attest the size and form of the persistent pulp; the diameter of its base is equal to the part of the crown of the tooth which is formed by the coarse and fine dentine. From the gradual thinning off, and final disappearance of these substances as they reach the base of the tooth, I conclude that they were both formed at the expense of the pulp. The fine tubes and cells must have been excavated in its peripheral layer for the reception of the hardening salts of the dense dentine, and the rest converted into the parallel series of medullary canals with their respective systems of calcigerous tubes, in a manner closely analogous to the development of the entire tooth of the Orycteropus. The coarser dentine of the tooth of the Megatherium differs, in fact, from the entire tooth of the *Orycteropus*, only in that the parallel medullary canals and their radiating calcigerous tubes are not separated from the contiguous canals by a distinct layer of caementum, and that the medullary canals anastomose at their peripheral extremities. The wide spaces (*e*. Pl. XXXI) indicate the thickness of the dental capsule by the ossification of which the exterior stratum of cement was formed. It was not until I knew the true structure of the tooth of the Megatherium, that I could comprehend the mode of its formation. The parallel layers of enamel in the Elephant's grinder are formed, as is well known, by membranous plates passing from the coronal end of the closed capsule towards the base of the tooth; but a certain extent of enamel can only thus be formed, and when the crown of the grinder has once protruded, and come into use, the enamel cannot be added to. The modification of the structure of the tooth of the Megatherium readily permits the uninterrupted and continuous formation of the dense substance which is analogous to the enamel of the Elephant's grinder.

With respect to the question of the respective affinities of the Megatherium to the Bradypodoid or Dasypodoid families, the result of this examination of the teeth speaks strongly for its closer relationship with the former group: the Megalonyx, Mylodon, and Scelidotherium, in like manner correspond in the structure of their teeth with the Sloth, and differ from the Armadillo.

If from a similarity of dental structure we may predicate a similarity of food, it may reasonably be conjectured that the leaves and soft succulent sprouts of trees may have been the staple diet of the Megatherioid quadrupeds, as of the existing Sloths. Their enormous claws, I conclude, from the fossorial character of the powerful mechanism by which they were worked, to have been employed, / not, as in the Sloths, to carry the animal to the food, but to bring the food within the reach of the animal, by uprooting the trees on which it grew.

In the remains of the Megatherium we have evidence of the framework of a quadruped equal to the task of undermining and hawling down the largest members of a tropical forest. In the latter operation it is obvious that the immediate application of the anterior extremities to the trunk of the tree would demand a corresponding fulcrum, to be effectual, and it is the necessity for an adequate basis of support and resistance to such an application of the fore-extremities which gives the explanation to the anomalous development of the pelvis, tail, and hinder extremities in the Megatherioid quadrupeds. No wonder, therefore, that their type of structure is so peculiar; for where shall we now find quadrupeds equal, like them, to the habitual task of uprooting trees for food?

DESCRIPTION OF FRAGMENTS OF BONES, AND OF OSSEOUS TESSELATED DERMAL COVERING OF LARGE EDENTATA

It is now determined that there once existed in South America, besides the Megatherium, the Megalonyx, and the allied genera described in the preceding pages of the present work, gigantic species of the order *Bruta* belonging to the Armadillo family, and defended, like the small existing representatives of that family, by a tesselated bony dermal covering. The largest known species of these extinct *Dasypodidae* is the *Glyptodon clavipes*, of which the armour and parts of the skeleton have been described by MM. Weiss and D'Alton in the *Berlin Transactions* for 1827 and 1834: and the generic and specific characters and name, with an account of the dental system, and bones of the extremities, were recorded in the *Geological Proceedings* for March 1839. It would seem that parts of the same, or a nearly allied gigantic species were described in the same year by M. Lund; under the name of *Hoplophorus*. Of the valuable and interesting discoveries of this able Naturalist I regret that I was not aware until the appearance of a notice of them in the *Comptes Rendus* for April, 1839.[64] Amongst the fragments of bony tesselated armour in Mr Darwin's collection are a few pieces which were found by him, associated with remains of Toxodon and Glossotherium near the Rio Negro in Banda Oriental.[65] These fragments, if we may judge from their thickness, must have belonged to an animal at least as / large as the *Glyptodon clavipes*; but the pattern differs in the greater equality of size of the component tesserae. The thickness of the largest fragment is one inch and a half, the tesserae vary in diameter from one inch to half an inch, and are separated by grooves about two lines in depth, and two in diameter. The pattern formed by the anastomosis of these grooves is an irregular net-work; the contour of the tesserae is either unevenly subcircular, hexagonal, pentagonal, or even four-sided; with the sides more or less unequal. In those portions of this armour, where one of the tesserae exceeds the contiguous ones in size, the imagination may readily conceive it to be the centre of a rosette, around which the smaller ones arrange themselves, but there is no regular system of rosettes, as in the portions of the dermal armour of the Glyptodon figured by Weiss, and those brought to England by Sir Woodbine Parish, in which the central piece is double the size of the marginal ones.

[64] An excellent translation of the description of the Brazilian fossils found by M. Lund, is published in the *Annals of Natural History*, July and August, 1839.

[65] At the distance of a few leagues from the locality here mentioned, other fragments were found by Mr Darwin; also near Santa Fé, in Entre Rios; also on the shores of the Laguna, near the Guardia del Monte, South of Buenos Ayres; also, according to the Jesuit Falkner, on the banks of the Tercero.

The portions of the tesselated bony dermal covering of a Dasypodoid quadruped, figured in Pl. XXXII, figs 5 and 4, of the natural size, were discovered folded round the middle and ungueal phalanges, figs 2 and 3, at Punta Alta, in Bahia Blanca, in an earthy bed interstratified with the conglomerate containing the remains of the fossil Edentals.

In one of these fragments, measuring six inches long by five broad, the tesserae are arranged in rosettes, and so closely correspond in size and pattern with the bony armour described by M. Lund, as characterizing his species, *Hoplophorus euphractus*, that I feel no hesitation in referring them to that animal. One of the pattern rosettes is figured at fig. 4, together with the thickness of the armour at this part, and the coarse tubulo-cellular structure of the bone. Another portion of dermal armour from the same locality, gives the pattern shown in fig. 5, formed by square or pentagonal tesserae, arranged in transverse rows; it is certain that this portion of armour belonged to the same animal as the preceding piece; and probably that it constituted part of the transverse dorsal bands of the *Hoplophorus*.

The middle and ungueal phalanx, as well as the portions of armour, are given of the natural size, in Pl. XXXII. The upper and outer surface of the phalanx, is shown in fig. 2. It is smooth and flat; joins the inner surface by a sharp edge, which runs along the upper and inner side of the bone; and passes by a gradual convexity to the under surface; the ridge corresponding with the base of the claw, is feebly developed at the under and lateral parts of the base of the claw. Below the double trochlear joint for the middle phalanx, there are two articular surfaces for two large sesamoid bones.

The middle phalanx corresponds in its small antero-posterior diameter and wedge-shape, with that of the great Glyptodon: but the terminal phalanx is longer and deeper, in proportion to its breadth. /

Among the collection of fossils from Punta Alta, in Bahia Blanca, there is an interesting fragment of the head of a gigantic animal of the Edentate order, including the glenoid cavity, and part of the zygomatic process of the left side. The articular surface for the lower jaw, exhibits, in its flatness, extent, and the absence of a posterior ridge, the well-marked characteristics of this part of the Edental structure. It measures two inches four lines in the transverse, and two inches two lines in the antero-posterior diameter. The commencement of the zygomatic process presents a vertical diameter of two inches, and a transverse diameter of eight lines at the thickest part. It is slightly concave at its lower border, and convex above. The small portion of the cranial parietes, which is preserved, exhibits the cellular structure consequent upon the great extension and development of the nasal air-sinuses: this condition of the cranial parietes, has already been noticed in the description of the more perfect skulls of the large extinct Edentata.

NOTICE OF FRAGMENTS OF MOLAR TEETH OF A

MASTODON

Of the remains of this gigantic extinct Pachyderm, observed by Mr Darwin at Santa Fé, in Entre Rios, and on the banks of the Tercero, the fragments of the teeth and portions of the skeleton which reached England, are not sufficient to lead to a determination of the species; but sufficiently prove it to have been nearly allied, if not identical, with the *Mastodon angustidens* of Cuvier, and unquestionably distinct from the *Mastodon giganteum* of the United States.

NOTICE OF THE REMAINS OF A SPECIES OF

EQUUS

Found associated with the extinct Edentals and Toxodon at Punta Alta, in Bahia Blanca, and with the Mastodon and Toxodon at Santa Fé, in Entre Rios

The first of these remains is a superior molar tooth of the right side; it was embedded in the quartz shingle, formed of pebbles strongly cemented together with calcareous matter, which adhered as closely to the tooth in question, as the corresponding matrix did to the associated fossil remains. The tooth was as completely fossilized as the remains of the Mylodon, Megatherium, and Scelidothere; and was so far decomposed, that in the attempt to detach the adherent matrix, it / became partially resolved into its component curved lamellae. Every point of comparison that could be established proved it to differ from the tooth of the common *Equus Caballus* only in a slight inferiority of size.

The second evidence of the co-existence of the horse with the extinct Mammals of the tertiary epoch of South America reposes on a more perfect tooth, likewise of the upper jaw, from the red argillaceous earth of the Pampas at Bajada de Santa Fé, in the Province of Entre Rios.[66]

This tooth is figured at Pl. XXXII, fig. 13 and 14, from which the anatomist can judge of its close correspondence with a middle molar of the left side of the upper jaw.

This tooth agreed so closely in colour and condition with the remains of the Mastodon and Toxodon from the same locality, that I have no doubt respecting the contemporaneous existence of the individual horse, of which it once formed part.

This evidence of the former existence of a genus, which, as regards South America, had become extinct, and has a second time been introduced into that Continent, is not one of the least interesting fruits of Mr Darwin's palaeontological discoveries.

DESCRIPTION OF REMAINS OF RODENTIA, INCLUDING THE JAWS AND TEETH OF AN EXTINCT SPECIES OF

CTENOMYS

The fragment of the upper jaw, figured in Pl. XXXII, fig. 6, exhibits the first and second molar *in situ*, and the socket of the third and fourth molar, of a Rodent, which by the form and number of the upper maxillary teeth is referable to the genus Ctenomys. The molars are a little larger, the longitudinal groove on their external surface is somewhat deeper, and the last molar is relatively wider than in the existing subterraneous species – the Tucutucu (*Ctenomys Brasiliensis*, Bl.) of whose habits so interesting an account is given in the description of the Mammalia of the present Collection (No. IV, p. 79). The form of the grinding surface of the first and second upper molar is shown below the fig. 6, and three views of the second grinder are given at figs 7, 8, and 9. The fragment of the lower jaw of the same fossil Rodent is figured at fig. 10 and 11. The long anterior incisor is relatively narrower than in the *Ctenomys Brasiliensis*. I have not had the means of

[66] Mr Darwin has more particularly described the circumstances of the embedment of this tooth in his *Journal of Researches*, p. 149, during the voyage of the *Beagle*.

comparing this fossil with the *Ctenomys Magellanicus*; but since it is probable that the *Ct. Magellanicus* may not be specifically different from the *Ct.* / *Brasiliensis*, it may be concluded that the present fossil is equally distinct from both.

The portion of the right hind-foot of the Rodent figured at fig. 12, includes the calcaneum, astragalus, cuboides, external and middle cuneiform bones, and the metatarsals and proximal phalanges of the toes corresponding with the three middle toes of five-toed quadrupeds. The metatarsals are chiefly remarkable for the well-developed double-trochlear articular surface, and intermediate ridge. These remains; as well as the jaws and teeth of the Ctenomys, were discovered at Monte Hermoso in Bahia Blanca.

In the same reddish earthy stratum of that locality, Mr Darwin discovered the decomposed molar of a Rodent, equalling in size, and closely resembling in the disposition of its oblique component laminae, the hinder molar of the Capybara (*Hydrochaerus*). The fossil differs, however, in the greater relative breadth of the component laminae.

I have, lastly, to notice the head of a femur, and some fragments of pelvic bones from the same formation which bear the same proportion to the tooth above alluded to, as subsists between the teeth and bones of the Capybara, and which are sufficient to prove that there once has existed in South America a species of the family *Caviidae*, as large as the present *Capybara*, but now apparently extinct.

This fact, together with the greater part of those which have been recorded in the foregoing pages of the present work, establishes the correspondence, in regard to the characteristic type, which exists between the present and extinct animals of the South American Continent: we have abundant evidence likewise of the greater number of generic and specific modifications of these fundamental types which the animals of a former epoch exhibited, and also of the vastly superior size which some of the species attained.

At the same time it has been shewn that some of the present laws of the geographical distribution of animals would not have been applicable to South America, at the period when the Megatherioids, Toxodon, and Macrauchenia existed: since the Horse, and according to M. Lund, the Antelope and the Hyaena, were then associated with those more strictly South American forms. The Horse, which, as regards the American continent, had once become extinct, has again been introduced, and now ranges in countless troops over the pampas and savannahs of the new world. If the small Opossums of South America had been in like manner imported into Europe, and were now established like the Squirrels and Dormice in the forests of France, an analogous case would exist to that of the Horse in South America, as the fossil Didelphys of Montmartre proves. /

With respect to the geological contemporaneity of the fossils collected by him, Mr Darwin subjoins the following observations:

'The remains of the following animals were embedded together at Punta Alta in Bahia Blanca: The *Megatherium Cuvierii, Megalonyx Jeffersonii, Mylodon Darwinii, Scelidotherium leptocephalum, Toxodon Platensis*(?) a Horse and a small Dasypodoid quadruped, mentioned p. [107] 141; at St Fé in Entre Rios, a Horse, a Mastodon, *Toxodon Platensis*, and some large animal with a tesselated osseous dermal covering; on the banks of the Tercero the Mastodon, Toxodon, and, according to the Jesuit Falkner, some animal with the same kind of covering; near the Rio Negro in Banda Oriental, the *Toxodon Platensis*, Glossotherium, and some animal with the same kind of covering. To these two latter animals the *Glyptodon clavipes*, described by Mr Owen in the *Geological Transactions*, may, from the locality where it was discovered, and from the similarity of the deposit which covers the greater part of Banda Oriental, almost certainly be added, as having been contemporaneous. From nearly the same reasons, it is probable that the Rodents found at Monte Hermoso in Bahia Blanca, co-existed with the several gigantic mammifers from Punta Alta. I have, also, shown in the Introduction, that the *Macrauchenia Patachonica*, must have been coeval, or nearly so, with the last mentioned animals. Although we have no evidence of the geological age of the deposits in some of the localities just specified, yet from the presence of the same fossil mammifers in others, of the age of which we have fair means of judging (in relation to the usual standard of comparison, of the amount of change in the specific forms of the invertebrate inhabitants of the sea), we may safely infer that *most* of the animals described in this volume, and likewise the Glyptodon, were strictly contemporaneous, and that *all* lived at about the same very recent period in the earth's history. Moreover, as some of the fossil animals, discovered in such extraordinary numbers by M. Lund in the caves of Brazil, are identical or closely related with some of those, which lately lived together in La Plata and Patagonia, a certain degree of light is thus thrown on the antiquity of the ancient Fauna of Brazil, which otherwise would have been left involved in complete darkness.'

Mammalia

DESCRIBED BY

GEORGE R. WATERHOUSE

WITH

A GEOGRAPHICAL INTRODUCTION

BY

CHARLES DARWIN

THE

ZOOLOGY

OF

THE VOYAGE OF H. M. S. BEAGLE,

UNDER THE COMMAND OF CAPTAIN FITZROY, R.N.,

DURING THE YEARS

1832 TO 1836.

PUBLISHED WITH THE APPROVAL OF
THE LORDS COMMISSIONERS OF HER MAJESTY'S TREASURY.

Edited and Superintended by
CHARLES DARWIN, ESQ. M.A. F.R.S. SEC. G.S.
NATURALIST TO THE EXPEDITION.

PART II.

MAMMALIA,

BY
GEORGE R. WATERHOUSE, ESQ.
CURATOR OF THE ZOOLOGICAL SOCIETY OF LONDON, ETC. ETC.

LONDON:
PUBLISHED BY SMITH, ELDER AND CO. 65, CORNHILL.
MDCCCXXXIX.

LIST OF PLATES

Plate XXXIV *continued*

teeth of upper jaw. Fig. 3. *c*. Molar teeth of lower jaw. Fig. 3. *d*. *Ramus* of lower jaw – natural size. Fig. 4. *a*. Skull of *Mus gracilipes*. Fig. 4. *b*. Molar teeth of upper jaw. Fig. 4. *c*. Molar teeth of lower jaw. Fig. 4. *d*. View of the under side of the tarsus. Fig. 5. *a*. First and second molar teeth of upper jaw of *Mus flavescens*. Fig. 5. *b*. Two posterior molar teeth of the lower jaw of *Mus flavescens*. Fig. 6. *a*. Molar teeth of the upper jaw of *Mus Magellanicus*. Fig. 6. *b*. Molar teeth of the lower jaw of *Mus Magellanicus*. Fig. 7. *a*. Skull of *Mus arenicola*. Fig. 7. *b*. Molar teeth of upper jaw. Fig. 7. *c*. Molar teeth of lower jaw. Fig. 7. *d*. *Ramus* of lower jaw. Fig. 8. *a*. Molar teeth of upper jaw of *Mus brachiotis*. Fig. 8. *b*. Two posterior molars of lower jaw. Fig. 9. *a*. Molar teeth of upper jaw of *Mus obscurus*. Fig. 9. *b*. Molar teeth of lower jaw of *Mus obscurus*. Fig. 10. *a*. *Ramus* of lower jaw of *Mus nasutus*. Fig. 11. *a*. Molar teeth of lower jaw of *Mus tumidus*. Fig. 12. *a*. *Ramus* of lower jaw of *Mus Braziliensis*. Fig. 13. *a*. Molar teeth of upper jaw of *Mus micropus*. Fig. 13. *b*. Molar teeth of lower jaw of *Mus micropus*. Fig. 14. *a*. *Ramus* of lower jaw of *Mus Galapagoensis*. Fig. 15. *a*. Molar teeth of upper jaw of *Mus griseo-flavus*. Fig. 15. *b*. Molar teeth of lower jaw of *Mus griseo-flavus*. Fig. 16. *a*. Molar teeth of upper jaw of *Mus xanthopygus*. Fig. 16. *b*. Molar teeth of lower jaw of *Mus xanthopygus*. Fig. 17. *a*. Molar teeth of upper jaw of *Mus Darwinii*. Fig. 17 *b*. Molar teeth of lower jaw of *Mus Darwinii*. Fig. 18. *a*. Molar teeth of upper jaw of *Mus Gouldii*. Fig. 18. *b*. Molar teeth of lower jaw of *Mus Gouldii*. Fig. 19. *a*. Molar teeth of upper jaw of *Mus insularis*. Fig. 19. *b*. Molar teeth of lower jaw of *Mus insularis*. Fig. 19. *c*. Portion of ramus of lower jaw. Fig. 20. *a*. Skull of *Reithrodon chinchilloides* – natural size. Fig. 20. *b*. Skull of *Reithrodon chinchilloides* viewed from beneath. Fig. 20. *c*. Skull of *Reithrodon chinchilloides*, side view. Fig. 20. *d*. *Ramus* of lower jaw – natural size. Fig. 20. *e*. Molar teeth of upper jaw. Fig. 20. *f*. Molar teeth of lower jaw. Fig. 21. *a*. Skull of *Reithrodon cuniculoïdes*, viewed from beneath. Fig. 21. *b*. Skull of *Reithrodon cuniculoïdes*, side view of fore part. Fig. 21. *c*. *Ramus* of lower pair. Fig. 22. *a*. Hinder part of ramus of lower jaw of *Abrocoma Bennettii*. Fig. 23. *a*.

[1] The palatine foramina are accidentally omitted – see description.

INDEX TO THE SPECIES

GEOGRAPHICAL INTRODUCTION

BY CHARLES DARWIN

The object of the present Introduction, is briefly to describe the principal localities, from which the Zoological specimens, collected during the voyage of the *Beagle*, were obtained. At the conclusion of this work, after each species has been separately examined and described, it will be more advantageous to incorporate any general remarks. The *Beagle* was employed for nearly five years out of England; of this time a very large proportion was spent in surveying the coasts of the Southern part of South America, and of the remainder, much was consumed in making long passages during her circumnavigation of the globe. Hence nearly the entire collection, especially of the animals belonging to the higher orders, was procured from this continent; to which, however, must be added the Galapagos Archipelago, a group of islands in the Pacific, but not far distant from the American coast. The localities may be briefly described under the following heads.

Brazil. This country presents an enormous area, supporting the most luxuriant productions of the intertropical regions. It is composed of primary formations, and may be considered as being hilly rather than mountainous. La Plata includes the several provinces bordering that great river; namely, Buenos Ayres, Banda Oriental, Santa Fé, Entre Rios, etc. My collections were chiefly made at Buenos Ayres, at Monte Video, the capital of Banda Oriental, and at Maldonado, a town in the same province, situated on the northern / shore, near the mouth of the estuary of the Plata. These countries consist either of an undulating surface, clothed with turf, or of perfectly level plains with enormous beds of thistles. Except on the banks of the rivers, trees nowhere grow; there are, however, thickets in some of the valleys, in the more hilly parts of Banda Oriental. During the winter and spring of this hemisphere, a considerable quantity of rain falls, and the plains of turf are then everywhere verdant; but in summer the country assumes a brown and parched appearance.

Bahia Blanca forms a large bay, in latitude 39°S. on a part of the coast, which falls within the territory of the province of Buenos Ayres, but which from its physical conditions would more properly be classed with Patagonia. The tertiary plains of Patagonia, extend from the Strait of Magellan to the Rio Negro, which is commonly assumed as their Northern boundary. This space of more than seven hundred miles in length, and in breadth reaching from the Cordillera to the Atlantic Ocean, is everywhere characterized by the dreary uniformity of its landscape. Nearly desert plains, composed of a thick bed of shingle, and often strewed over with sea-shells (plainly indicating that the land has been covered within a recent period by the sea), are but rarely interrupted by hills of porphyry, and other crystalline rocks. The plains support scattered tufts of wiry grass, and stunted bushes; whilst in the broad flat-bottomed valleys, dwarf thorn-bearing trees, barely ornamented with the scantiest foliage, sometimes unite into thickets; and here the few feathered inhabitants of these sterile regions resort. There is an extreme scarcity of water; and where it is found, especially if in lakes, it is generally as salt as brine. The sky in summer is cloudless, and the heat in consequence, considerable; whereas the frosts of winter are, sometimes, severe. The principal localities visited by the *Beagle*, were the Rio Negro, in latitude 41°S., Port Desire, Port St Julian, and Santa Cruz. At the latter place, a party, under the command of Captain FitzRoy, followed up the river in boats, to within a few miles of the Cordillera; and an opportunity was thus afforded of verifying the nature of the country in its entire breadth. At the Rio Negro the plains are much more thickly covered with bushes (chiefly acacias), than in any other part of Patagonia.

Tierra del Fuego may be supposed to include all the broken land south of a line joining the opposite mouths of the Strait of Magellan. The land is mountainous, / and may be aptly compared to a lofty chain, partly submerged in the sea; bays and channels occupying the position of valleys. The Eastern side almost exclusively consists of clay-slate; the Western, of primary, and various plutonic formations. The mountains, from the water's edge, to within a short distance of the lower limit of perpetual snow, are everywhere (excepting on the exposed western shores) concealed by an impervious forest, the trees of which do not periodically shed their leaves. On the East coast, the outline of the land shows that tertiary formations, like those of Patagonia, extend south of the Strait of Magellan; but with the exception of this part, it is rare to find even a small space of level ground; and where such occurs, a thick bed of peat invariably covers the surface. The climate is of that kind which has been denominated insular: the winters are far from being excessively cold, whilst the summers are gloomy, boisterous, and seldom cheered by the rays of the sun. In all seasons, a large quantity of rain falls. Hence, from the physical conditions of Tierra del Fuego, all the land animals must live either on the sea beach (and in this class the Aborigines may be included) or within the humid and entangled forests.

The Falkland Islands are situated in the same latitude as the Eastern entrance of the Strait of Magellan, and about 270 miles East of it. The climate is nearly the same as in Tierra del Fuego, but the surface of the land, instead of being as there, concealed by one great forest, does not support a single tree. We see on every side a withered and coarse herbage, with a few low bushes, which spring from the peaty soil of an undulating moorland. Scattered hills, and a central range of quartz rock, protrude through formations of clay-slate and sand-stone (belonging to the Silurian epoch) which compose the lower country.

The structure of the west coast of South America, from the Strait of Magellan northward to latitude 38°, in its greater part (as far north as Chiloe), is very similar to that of Tierra del Fuego. The climate likewise is similar – being gloomy, boisterous, and extremely humid; and, consequently, the land is concealed by an almost impenetrable forest. In the northern part of this region, the temperature of course is considerably higher than near the Strait of Magellan; but nevertheless it is much less so, than might have been anticipated from so / great a change in latitude. Hence, although the vegetation of this northern district presents a marked difference when compared with that of the southern; yet the zoology in many respects has, like the general aspect of the landscape, a very uniform character. The specimens were chiefly collected from the Peninsula of Tres Montes, the Chonos Archipelago (from latitude 46° to 43° 30′), Chiloe with the adjoining islets, and Valdivia. The contrast between the physical conditions and productions of the East and West coasts of this part of South America is very remarkable. On one side of the Cordillera, great heavy clouds are driven along by the western gales in unbroken sheets, and the indented land is clothed with thick forests; whilst on the other side of this great range, a bright sky, with a clear and dry atmosphere, extends over wide and desolate plains.

Chile in the neighbourhood of Concepcion (latitude 36° 42′S.) may be called a fertile land; for it is diversified with fine woods, pasturage, and cultivated fields. But towards the more central districts (near Valparaiso and Santiago) although by the aid of irrigation, the soil in the valleys yields a most abundant return, yet the appearance of the hills, thinly scattered with various kinds of bushes and cylindrical Opuntias, bespeaks an arid climate. In winter, rain is copious, but during a long summer of from six to eight months, a shower never moistens the parched soil. The country has a very alpine character, and is traversed by several chains of mountains extending parallel to the Andes. These ranges include between them level basins, which appear once to have formed the beds of ancient channels and bays, such as those now intersecting the land further to the south. North of the neighbourhood of Valparaiso, the climate rapidly becomes more and more arid, and the land in proportion desert. Beyond the valley of Coquimbo (latitude 30°) it is scarcely habitable, excepting in the valleys of Guasco, Copiapo, and Paposa, which owe their entire fertility to the system of irrigation, invented by the aboriginal Indians and followed by the Spanish colonists. Northward of these places, the absolute desert of Atacama forms a complete barrier, and eastward, the snow-clad chain of the Cordillera separates the Zoological province of Chile, from that of the wide plains which extend on the other side of the Andes.

The last district which it is at all necessary for me to mention here, is that / of the Galapagos Archipelago, situated under the Equator, and between five and six hundred miles West of the coast of America. These islands are entirely volcanic in their composition; and on two of them the volcanic forces have within late years been seen in activity. There are five principal islands, and several smaller ones: they cover a space of 2° 10′ in latitude, and 2° 35′ in longitude. The climate, for an equatorial region, is far from being excessively hot: it is extremely dry; and although the sky is often clouded, rain seldom falls, excepting during one short season, and then its quantity is variable. Hence, in the lower part of these islands, even the more ancient streams of lava (the recent ones still remaining naked and glossy) are clothed only with thin and nearly leafless bushes. At an elevation of 1,200 feet, and upwards, the land receives the moisture condensed from the clouds, which are drifted by the trade wind over this part of the ocean at an inconsiderable height. In consequence of this, the upper and central part of each island supports a green and thriving vegetation; but from some cause, not very easily explained, it is much less frequented, than the lower and rocky districts are, by the feathered inhabitants of this archipelago.

By a reference to the localities here described, it is hoped that the reader will obtain some general idea of the nature of the different countries inhabited by the several animals, which will be described in the following sheets.

The vertebrate animals in my collection have been presented to the following museums: the Mammalia and Birds to the Zoological Society; the Fishes to the Cambridge Philosophical Society; and the Reptiles, when described, will be deposited in the British Museum. For the care and preservation of all these and other specimens, during the long interval of time between their arrival in this country and my return, I am deeply indebted to the kindness of the Rev. Professor Henslow of Cambridge. With respect to the gentlemen, who have undertaken the several departments of this publication, I hope they will permit me here to express the great personal obligation which I feel towards them, and likewise my admiration at the disinterested zeal which has induced them thus to bestow their time and talents for the good of Science. /

PLATE I

Desmodus D'Orbignyi

MAMMALIA

BY GEORGE WATERHOUSE

FAMILY: PHYLLOSTOMIDAE

DESMODUS D'ORBIGNYI

PLATE I Skull, teeth, etc. Pl. XXXV, fig. 1

D. pilis nitidis adpressis; corpore suprà fusco, pilis ad basin albis; gulá abdomineque cinerescenti-albis; nasûs prosthemate parvulo bifido

Description. The fur of this Bat is glossy and has a silk-like appearance; that on the top of the head, sides of the face, and the whole of the upper parts of the body, is of a deep brown colour; all the hairs on these parts, however, are white at the base. The flanks, interfemoral membrane, and the arms, are also covered on their upper side with brown hairs. On the lower part of the sides of the face, and the whole of the under parts of the body, the hairs are of an ashy-white colour. The membrane of the wing is brownish. The ears are of moderate size, and somewhat pointed; externally they are covered with minute brown hairs, and internally with white. The tragus is also covered with white hairs; it is of a narrow form, pointed at the tip, and has a small acute process in the middle of the outer margin. The nose-leaf is pierced by the nostrils, which diverge posteriorly, and is so deeply cleft on its hinder margin, that it may be compared to two small leaflets joined side by side near their bases. These leaflets, unlike the nose-leaf of the Phyllostomina, lie horizontally on the nose to which they are attached throughout, a slight ridge only indicating their margin. Around the posterior part of the nose-leaf there is a considerable naked space, in which two small hollows are observable, situated one each side, and close to the / nose-leaf; and, at a short distance behind the nose-leaf, this naked membrane is slightly elevated, and forms a transverse fleshy tubercle.

	Inches	*Lines*		*Inches*	*Lines*
Length of head and body	3	3	Length of tarsus (claw included)	0	8⅓
interfemoral membrane	0	3½	ear	0	4
the antibrachium	2	2	tragus	0	3
thumb (claw included)	0	8	nose-leaf	0	2¹⁄₁
tibia	0	10	Expanse of the wings	12	8

Habitat, Coquimbo, Chile. (*May*)

'The Vampire Bat,' says Mr Darwin in his MS. notes upon the present species, 'is often the cause of much trouble, by biting the horses on their withers. The injury is generally not so much owing to the loss of blood, as to the inflammation which the pressure of the saddle afterwards produces. The whole circumstance has lately been doubted in England; I was therefore fortunate in being present when one was actually caught on a horse's back. We were bivouacking late one evening near Coquimbo, in Chile, when my servant, noticing that one of the horses was very restive, went to see what was the matter, and fancying he could distinguish something, suddenly put his hand on the beast's withers, and secured the Vampire. In the morning, the spot where the bite had been inflicted was easily distinguished from being slightly swollen and bloody. The third day afterwards we rode the horse, without any ill effects.

Before the introduction of the domesticated quadrupeds, this Vampire Bat probably preyed on the guanaco, or vicugna, for these, together with the puma, and man, were the only terrestrial mammalia of large size, which formerly inhabited the northern part of Chile. This species must be unknown, or very uncommon in Central Chile, since Molina, who lived in that part, says (*Compendio de la Historia del Reyno de Chile*, vol. i, p. 301) 'that no blood-sucking species is found in this province'.

It is interesting to find that the structure of this animal is in perfect accordance with the habits as above detailed by Mr Darwin. Among other points, the total absence of true molars, and consequent want of the power of masticating food is the most remarkable. On the other hand we find the canines and incisors perfectly fitted for inflicting a wound such as described, while the small size of the interfemoral membrane (giving freedom to the motions of the legs), together with the unusually large size of the thumb and claw, would enable this Bat, as I should imagine, to fix itself with great security to the body of the horse.

I have named this species after M. d'Orbigny, who has added so much to / our information on the zoölogical productions of South America. The *Edostoma cinerea*[1] of that author has evidently a close affinity to the animal here described, and differs chiefly (judging from the drawing published in his work) in the larger size of the ears, in having the nose-leaf free, and the surrounding membrane free and elevated.

As M. d'Orbigny has not yet published the character of his genus *Edostoma*, his figure is my only guide, and in this figure I find the dentition agreeing both with that of the present species, and that of the genus *Desmodus* of Prince Maximilian – as would appear from the published descriptions, and figure given by M. de Blainville.[2] The points of distinction between M. d'Orbigny's animal and the species here described, are not, in my opinion, of sufficient importance to constitute generic characters, I have, therefore, retained the name of Desmodus.

It is desireable perhaps to separate the Blood-sucking Bats from the Insectivorous species, and place them between the latter group and the *Pteropina* (with which they agree in the large size of the thumb and the rudimentary interfemoral membrane), under a sectional name, which I propose to call *Haematophilini*.

1. PHYLLOSTOMA GRAYI

PLATE II

P. fusco-cinereum; nasûs prosthemate lanceolato; auribus mediocribus, trago basin versus extùs unidentato; caudâ gracillimâ, brevi, et membranâ interfemorali inclusâ; verrucâ complanatâ ad apicem menti, verrucis parvulis circumdatâ

Description. This Phyllostoma agrees with the species described by

[1] *Voy. Amer. Merid.*, t. 8.

[2] See his memoir 'Sur quelques anomalies du système dentaire dans les mammifères,' published in the *Annales Françaises et Etrangères d'Anatomie et de Physiologie*, No. 6, pl. IX, fig. 2.

Mr J. Gray[3] under the name of Childreni, in having on the lower lip 'an half ovate group of crowded warts', but is of a much smaller size, and differs also in colour.

The number of teeth are as follows: incisors 4/4; canines 2/2; molars $\frac{5-5}{5-5}$=32. The intermediate pair of incisors of the upper jaw are large, compressed, and have their apices rounded; the lateral pair are so minute, that they are scarcely visible without the assistance of a lens: the four incisors of the / lower jaw, are somewhat crowded, the intermediate pair are slightly larger than the lateral; they are all deeply notched, and broad at the apex. The cerebral portion of the skull is much arched and the anterior portion is depressed. The zygomatic arch is imperfect; see Pl. XXXV, figs 2. The nose-leaf is lanceolate, and of moderate size; the ears are also of moderate size; they are rounded at the tip and emarginated on their exterior edge: the tragus is elongated, and suddenly attenuated towards the apex; the outer margin is deeply notched towards the base, and very obscurely crenulated above this notch. The interfemoral membrane is of moderate extent, and emarginated posteriorly. The tail, which is very slender, is entirely enclosed by the interfemoral membrane, and the visible portion appears to consist of but two joints, which together, measure about two and a half lines in length. The basal half of the thumb is enclosed in membrane. The fur is soft and rather long. The general tint of the upper and under parts of the body is brownish-ash; the hairs on the neck and on the whole of the back are grey at the base, then white, or nearly so, brownish-ash near the tip, and whitish at the tip. On the belly the hairs are nearly of an uniform brown-ash colour, their apices only being whitish. The ears, nose-leaf, and membrane of the wings, are of a sooty-black hue.

	Inches	*Lines*		*Inches*	*Lines*
Length of head and body	2	0	Length of ear	0	7
antibrachium	1	4⅓	nose-leaf	0	3⅓
thumb (claw included)	0	5½	Expanse of the wings	10	0
tibia	0	7			

Habitat, Pernambuco, Brazil. (*August*)

'This species appeared to be common at Pernambuco (five degrees north of Bahia). Upon entering an old lime-kiln in the middle of the day, I disturbed a considerable number of them: they did not seem to

[3] *Magazine of Zoology and Botany*, No. 12.

PLATE II
Phyllostoma Grayi

be much incommoded by the light, and their habitation was much less dark than that usually frequented as a sleeping place by these animals.' D.

I have named this species after Mr John Gray, the author of several extensive memoirs on the order to which it belongs, and to whom I am indebted for valuable assistance whilst comparing this and other species with those contained in the collection of the British Museum.

2. PHYLLOSTOMA PERSPICILLATUM

I find in Mr Darwin's collection, a bat agreeing with the description of M. / Geoffroy Saint Hilaire,[4] under the above name, with the exception of a slight difference in the dimensions; I will, therefore, add those of the present specimen, which is a female. It may be observed, that in the animal before me, the tragus of the ear is pointed, and not bifid at the apex, as represented in plate xi of the work quoted.

	Inches	*Lines*		*Inches*	*Lines*
Length of head and body	4	0	Length of tragus	0	3
antibrachium	2	7	tibia	1	0
nose-leaf	0	5	Expansion of the wings	16	8
ear	0	8½			

'This bat was caught at Bahia (latitude 13°S.) on the coast of Brazil, in consequence of its having flown into a room where there was a light. I scarcely ever saw an animal so tenacious of life.' D.

FAMILY: VESPERTILIONIDAE

VESPERTILIO CHILOENSIS

PLATE III

V. fuscus: auribus mediocribus; trago elongato, angusto, apicem versus attenuato; fronte concavo; rostro obtuso; caudâ ad apicem extremum liberâ

Description. In size and colouring, this Bat very closely resembles the

[4] *Annales des Muséum d'Histoire Naturelle*, tom. xv, p. 176.

PLATE III

Vespertilio Chiloensis

Vespertilio Pipistrellus of Europe; the wings, however, are considerably broader in proportion; the antibrachium, tibia, and tail, are each of them longer; the tragus of the ear is also longer, and narrower.

The muzzle is short and obtuse, and furnished on each side with numerous hairs, which, when compared with those of other parts, are of a more harsh nature. The nose is naked at the apex. The forehead is concave. The ears are narrow, and somewhat pointed, emarginated externally, and have about four transverse rugae: the tragus is elongated, narrow, and pointed, and has the outer margin very obscurely crenulated. On the chin there is a small wart, from which spring several stiffish hairs. The tail is about equal to the body in length, and has the extreme tip free. The fur is moderately / long, and of an uniform rich brown colour, and extends on to the base of the interfemoral membrane above and below; the remainder of this membrane is bare, and, together with that of the wings, of a black colour.

	Inches	*Lines*		*Inches*	*Lines*
Length of the head and body	1	8	Length of the tragus	0	3⅓
tail	1	3½	the antibrachium	1	5½
Expanse of the wings	8	3	the thumb (claw included)	0	2½
Length of the ear	0	5½	the tibia	0	6¾

Habitat, Chiloe. (*January*)

'This specimen was given me by Lieutenant Sulivan, who obtained it among the islets on the Eastern side of Chiloe. It is not, I believe, common, nor do the humid and impervious forests of that island appear a congenial habitation for members of this family. It must, however, be observed, that even in Tierra del Fuego, where the climate is still less hospitable, and where the number of insects is surprisingly small, I saw one of these animals on the wing.' D.

FAMILY: NOCTILIONIDAE

DYSOPES NASUTUS

Molossus nasutus *Spix*, Simiarum et Vespertilionum. Braziliensium species novae.
Nyctinomus Braziliensis, *Geoffroy*, *Annales des Sciences Naturelles*, tom. i, p. 337, pl. 22.

Of this species I find three specimens in Mr Darwin's collection – 'It is remarkable,' says Mr Darwin, 'for its wide geographical range. I obtained specimens at Maldonado, on the northern bank of the Plata, where it was exceedingly numerous in the attics of old houses, and likewise at Valparaiso in Chile. Molina (vol. i, p. 301) says another species is found in Chile, of the same size and figure, but of a more orange (*naranjado*) colour.'

Upon comparing the dimensions of several specimens of this species with those given by Temminck in his *Monographie sur le Genre Molosse*, I find that they vary very considerably; I shall therefore be adding some little to the history of the species, by giving the dimensions of those now before me, together with the sexes of the specimens measured, and their localities. In all these specimens there is a series of pointed tubercles along the upper margin of the ears, a character which M. Temminck has omitted to notice. They vary slightly / in the intensity of their colouring, but among those brought from Chile I do not perceive any agreeing with that species, or variety, mentioned by Molina as approaching to an orange colour. All the specimens whose dimensions are here given, are preserved in spirit. Two of them are from Maldonado brought by Mr Darwin; three were collected in Hayti by Mr J. Hearne, and one is from Chile, whence it was brought by Mr H. Cuming.

	From Chile ♀		Hayti ♀		Hayti ♂		Hayti ♂		Maldonado ♀		Maldonado ♀	
	Inches	*Lines*	*Inches*	*Lines*	*Inches*	*Lines*	*Inches*	*Lines*	*Inches*	*Lines*	*Inches*	*Lines*
Length of head and body	2	3	1	11	2	0	2	0½	2	6	2	6
Length of tail	1	1½	1	2	1	2	1	1½	1	1	1	2
Length of free portion of tail	0	6½	0	5	0	6½	0	5⅓	0	8¾	0	8½
Expanse of wings	10	3	9	3	9	8	9	0	10	6	10	2
Length of antibrachium	1	7	1	6	1	6½	1	6	1	8	1	9
Length of ears	0	5	0	4½	0	4¾	0	4½	0	5⅓	0	5½
Width of ears	0	7	0	6	0	6	0	6	0	7	0	7
Length from nose to eye	0	3½	0	3	0	3¾	0	3	0	3½	0	3½

In all specimens examined by me, there are two incisors in the upper jaw, and four in the lower, they would therefore, according to M. Temminck, be adult.

FAMILY: CARNIVORA

1. CANIS ANTARCTICUS

PLATE IV

Antarctic Wolf, *Pennant, History of Quadrupeds*, vol. i, p. 257, sp. 165.
Canis Antarcticus, *Shaw, Gen. Zool.*, vol. i, pt 2, p. 331.
Canis Antarcticus, *Desm., Mamm.*, p. 199.

C. suprà sordidè fulvescenti-brunneus, pilis ad apicem nigris; lateribus, corporeque subtùs, sordidè flavescenti-fuscis; capite, auribusque extùs, fusco nigroque adspersis; artubus flavescenti-fulvis; labiis, gulâ, abdomine imo, femoribusque intùs, sordidè albis; caudâ ad basin concolore cum corpore, deìn nigrâ, apice albo

Description. This animal is considerably larger than the common fox (*Canis Vulpes*, Auct.), and stouter in its proportions, and, in fact, appears to be intermediate between the ordinary foxes and the wolves. The tail is much / smaller and less bushy than in the former animals. The contour of the head is wolf-like; the legs, however, are shorter than in the true wolves; and the tail is white at the apex, a character common in the foxes.

The fur of the Antarctic Fox is moderately long, and the under fur is not very abundant, especially as compared with that of the *C. magellanicus*. This under fur is of a pale brown colour; the apical portion of each hair is yellowish; the longer hairs are black at the apex, brown at the base, and annulated with white towards the apex. In many of these hairs the subapical pale ring is wanting. On the chest and belly the hairs are of a pale dirty yellow colour, grey-white at the base, and black at the apex. On the hinder part of the belly the hairs are almost of an uniform dirty white. The space around the angle of the mouth, the upper lip, and the whole of the throat, are white. The chin is brown-white, or brownish. The basal half of the tail is of the same colour as the body, and the hairs are of the same texture; on the apical half of the tail they are of a harsher or less woolly nature, of a black colour at the apex, and brownish at the base; those at the extreme point are totally white. The legs are almost of an uniform fulvous colour; the feet are of a somewhat paler hue;

PLATE IV

Canis antarcticus

the hairs on the under side of the hinder feet are brownish, and the external and posterior parts of the tibiae are suffused with the same tint. The hairs on the head are grizzled with black and fulvous; the former of these colours is somewhat conspicuous, excepting in the region of the eyes, where the fulvous or yellowish tint prevails. The muzzle is scarcely of so dark a hue as the crown of the head. The ears are furnished internally with long white hairs, externally the hairs are yellowish, with their apices black; the latter colour is more conspicuous towards the tip of the ear. The sides of the neck near the ear are of a rich fulvous hue.

	Inches	*Lines*
Length from nose to root of tail	36	0
from tip of nose to ear	7	3
of tail (hair included)	13	0
Length of ear	2	9
Height of body at shoulders	15	0

Habitat, Falkland Islands.

'Three specimens of this animal were brought to England by Captain FitzRoy; from one of which, the above drawing and description has been made. The earliest notice I can find of this animal is by Pernety,[5] during Bougainville's voyage, which was undertaken in 1764, for the purpose of colonizing these islands. The strange familiarity of its manner seems to have excited the fears of some of / the seamen in Commodore Byron's voyage (in 1765) in rather a ludicrous manner. Byron says that seals were not the only dangerous animals that they found, 'for the master having been sent out one day to sound the coast upon the south shore, reported at his return that four creatures of great fierceness, resembling wolves, ran up to their bellies in the water to attack the people in his boat, and that as they happened to have no fire-arms with them, they had immediately put the boat off in deep water'. Byron adds that, 'When any of these creatures got sight of our people, though at ever so great a distance, they ran directly at them; and no less than five of them were killed this day. They were always called wolves by the ship's company, but, except in their size, and the shape of the tail, I think they bore a greater resemblance to a fox. They are as big as a middle-sized mastiff, and their fangs are remarkably long and sharp. There

[5] *Journal Historique d'un Voyage fait aux Iles Malouines*, tom. ii, p. 459.

are great numbers of them upon this coast, though it is not perhaps easy to guess how they first came hither; for these islands are at least one hundred leagues distant from the main. They burrow in the ground like a fox, and we have frequently seen pieces of seals which they have mangled, and the skins of penguins lie scattered about the mouths of their holes. To get rid of these creatures, our people set fire to the grass, so that the country was in a blaze as far as the eye could reach, for several days, and we could see them running in great numbers to seek other quarters.

The habits of these animals remain nearly the same to the present day, although their numbers have been greatly decreased by the singular facility with which they are destroyed. I was assured by several of the Spanish countrymen, who are employed in hunting the cattle which have run wild on these islands, that they have repeatedly killed them by means of a knife held in one hand, and a piece of meat to tempt them to approach, in the other. They range over the whole island, but perhaps are most numerous near the coast; in the inland parts they must subsist almost exclusively on the upland geese (*Anser leucopterus*), which, from fear of them, like the eider-ducks of Iceland, build only on the small outlying islets. These wolves do not go in packs; they wander about by day, but more commonly in the evening; they burrow holes; are generally very silent, excepting during the breeding season, when they utter cries, which were described to me as resembling those of the *Canis Azarae*. Spaniards and half-cast Indians, from several districts of the southern portions of South America, have visited these islands, and they all declare that the wolf is not found on the mainland; the sealers likewise say it does not occur on Georgia, Sandwich Land, or the other islands in the Antarctic ocean. I entertain, therefore, no doubt, that the *Canis antarcticus* is peculiar to this archipelago. It is found both on East and West Falkland, as might have been inferred from the accounts given by Bougainville and Byron, who visited different islands; I state this particularly, because the contrary has been asserted. I was / assured by Mr Low, an intelligent sealer, who has long frequented these islands, that the wolves of West Falkland are invariably smaller and of a redder colour than those from the Eastern island; and this account was corroborated by the officers of the *Adventure*, employed in surveying the archipelago. Mr Gray, of the British Museum, had the kindness to compare in my presence the specimens deposited there by Captain FiztRoy, but he could not detect any essential difference between them. The number of these animals during the last fifty years must have been greatly reduced; already they are entirely banished from that half of East Falkland which lies East of the head of St Salvador Bay and Berkeley Sound; and it cannot, I think, be doubted, that as these islands are now becoming colonized, before the paper is decayed on which this animal has been figured, it will be ranked among those species which have perished from the face of the earth.' D

2. CANIS MAGELLANICUS

PLATE V

Canis Magellanicus, *Gray, Proceedings of the Zoological Society of London*, part iv, 1836, p. 88.

Vulpes Magellanica, *Gray, Magazine of Natural History*, New Series, 1837, vol. i, p. 578.

C. *suprà albo nigroque variegatus; lateribus fulvescente fuscoque lavatis; capite fusco-flavo et albescente adsperso; rostro supernè obscuriore; auribus, artubusque extùs flavescenti-rufis; corpore subtùs sordidè flavescenti-albo; pectore fulvo lavato; mento fuscescente; caudâ fulvescenti-fuscâ, pilis ad apicem nigris, subtùs pallidiore; plagâ supernè prope basin caudae, hujusque apice nigris*

Description. This species is considerably larger than the European fox; its form is more bulky, the limbs are shorter and stouter in proportion, the ears are smaller and the tail is more bushy. The fur is long, thick, and loose. The under fur is very long, abundant, and of a woolly texture. The back is mottled with black and white, the former of these colours being predominant; the hairs on this part are grey at the base, there is then a considerable space of a pale, or whitish brown colour; next follows a broad white ring, beyond which the hairs are black. On the sides of the body the hairs are coloured in the same way, excepting that the white portion is more extended, and is followed by a rich yellow-brown, shaded into black as it approaches the apex of each hair. Hence the general hue of the sides of the body is paler than that of the back, the brown and white tints being the more conspicuous. / The hairs of the head are annulated with white, and fulvous, and are black

PLATE V
Canis Magellanicus

at the tip; the two former colours are most conspicuous. The chin is brownish. The lower part of the cheeks, the throat, and the under parts of the body, are of a dirty yellowish white colour, inclining to buff in certain parts, especially on the lower part of the neck and chest. The limbs are of a rich deep fulvous, or yellowish rust colour externally; the feet and inner sides of the legs are of a paler hue. On the hinder legs externally, above the heel, is a patch of bright rust colour; such is also the colour of the ears externally, and likewise of that portion of the neck behind the ears. Internally the ears are furnished with long yellowish white hairs. The tail is long and very bushy; at its base the hairs are rusty white, towards the middle they are of a paleish rust colour, and at the apex they are black; there is also a black patch on the upper part towards the base. The hairs of the tail beneath are almost entirely of an uniform rusty white colour, those on the upper side are all tipped with black

	Inches	*Lines*		*Inches*	*Lines*
Length from nose to root of tail	31	0	Length of ear	2	0
from nose to base of ear	6	9	Height of body at the shoulders	14	6
of tail (hair included)	17	0			

Habitat, Chile. (*June*)

'This animal was first brought to Europe by Captain Philip P. King, who obtained it at Port Famine in Tierra del Fuego, where it is common. My specimen was obtained in the valley of Copiapó in the northern part of Chile. The Magellanic fox, therefore, has a range on the western coast of at least 1,600 miles, from the humid and entangled forests of Tierra del Fuego, to the almost absolutely desert country of northern Chile. In La Plata, on the Atlantic side of the continent, I believe it is not found.[6] It is mentioned by Molina in his account of the animals of Chile,[7] under the name of Culpeu, which he supposes to be derived from the Indian word 'culpem', signifying madness; for this animal, when it sees a man, runs towards him, and

[6] Azara has not described this animal, which circumstance alone would render it probable that it is not an inhabitant of Paraguay or La Plata. The two Foxes mentioned by him are the Aguará-guaza (*Canis jubatus*, Auct.), a very large kind of fox (a strangely exaggerated description of this animal is given by Falkner) of which I could not obtain a specimen; and the Aguará-chay, or *Canis Azarae*.

[7] Molina, *Compendio de la Historia del Reyno de Chile*, vol. i, pp. 330 and 332.

standing at the distance of a few yards, looks at him attentively. He adds, although great numbers are killed, they do not leave off this habit. Molina states that he has repeatedly been a witness of this, and I received nearly similar accounts from several of the inhabitants of Chile: yet I must observe, that the people of the farm-house, where my specimen was killed / (after it, together with its female, had destroyed nearly two hundred fowls) bitterly complained of its craftiness. From this bold curiosity in the disposition of the Culpeu, Molina thought that it was the same animal as that described by Byron at the Falkland Islands, but we now know that they are different. The Culpeu burrows holes under ground, often wanders about by day, is very strong and fleet. When riding one day in the valley of Copiapó, accompanied by a half-bred greyhound, I happened to come across one of these foxes; and although the ground was, in the first part of the chase, level, it soon entirely distanced its pursuer. Whilst running, it barked so like a dog, that until it had run some way ahead of the greyhound, I could not tell from which animal the noise proceeded. After the Culpeu had reached the mountains, it made a sudden bend from its course, and returned in a nearly parallel line, but at the base of a steep cliff of rock; it then quietly seated itself on its haunches, and seemed to listen with much satisfaction to the dog, which was running the scent on the mountain side, above its head.' D.

3. CANIS FULVIPES

PLATE VI

Canis fulvipes, *Martin, Proceedings of the Zoological Society of London*, 1837, p. 11.

C. *suprà niger, albo adspersus, capite lateribusque fuscis, sordidè albo nigroque adspersis; rostro superiore, mentoque fusco-nigricantibus; gulâ, labiis superioribus, femoribusque ad partem anteriorem, sordidè albis; pectore abdomineque fuscescentibus; auribus externè rufo-castaneis; brachiis internè, tarsis, digitisque fuscescenti-fulvis; artubus posticis extùs supra calcem fusco-nigrescentibus; caudae colore ad basin ut in corpore, apice nigro*

Description. This species is considerably less than the common European fox (*Canis Vulpes*, Auct.), its weight probably would scarcely exceed half that of the latter animal. The form of the body is stout, the limbs are short and rather slender; the head is also short, and the muzzle is pointed; the ears are of moderate size. The tail is about equal to half the whole length of the body, head included; and compared with that of ordinary foxes, is much less bushy, especially at the base. The general hue of this animal is very dark; the fur is rather short, and harsh to the touch; the under fur is abundant, and of a woolly texture. On the back, all the hairs are of a deep brown colour, annulated with white near the apex, and black at the apex. When the fur is / in its ordinary position, the brown colour is not seen, and the black and white produce a grizzled appearance; the black colour, however, predominates. On the sides of the body each hair is greyish at the base, then pale brown, near the apex annulated with white, and at the apex black: the three last mentioned colours are exhibited in about equal proportions (the fur being in its natural position) over the haunches and shoulders, but between these two parts, the brown and white colours are the more conspicuous, the hairs of the head are coloured in the same way as those of the sides of the body, excepting that the brown portion of each hair, is replaced by rusty brown, which gives a rufous hue to this part. The muzzle and chin are of a sooty brown colour. A dirty white patch is observable on each side of the muzzle at the apex; and this colour is extended along the margin of the upper lip on to the lower part of the cheeks, and over the whole of the throat; all the hairs in these parts (with the exception of those on the lips) being of a deep brownish grey colour, with their apical portions only, white. The ears are covered internally with long yellowish white hairs; towards, and on the margin of the ears externally, the hairs are of a buff colour, on the remaining portion of the ears, and on the sides of the neck, they are of a reddish chestnut hue. The hairs of the under parts of the body are brown, those near the hinder legs, and between them, are of a dirty white colour at the apex; towards the rump they are of a yellowish brown colour. The hairs of the tail are brown, black at the apex, and annulated with white near the apex; on the apical portion the hairs are black, and brown at the base. The fore legs are of a brown colour externally, internally they are of a brownish fulvous hue; such is also the colour of the feet. The fore part of the posterior legs is whitish, and there is a large blackish patch on the outer side, and extending around the posterior part, above the heel.

PLATE VI

Canis fulvipes

	Inches	*Lines*		*Inches*	*Lines*
Length from nose to root of tail	24	0	Length of tail (hair included)	10	0
from nose to base of ear	.0	$4\frac{3}{4}$	of ear	2	$3\frac{1}{2}$
			Height of body at shoulders	10	6

Habitat, Chiloe. (*December*)

'I killed this animal on the sea-beach, at the southern point of the island; it is considered extremely rare in the northern and inhabited districts. Molina mentions this fox, which he falsely considered as the *C. lagopus*, under the name of the *Payne Gurú*, and he adds, that in the Archipelago of Chiloe, it is found of a black colour. From this circumstance I am induced to believe that the species is confined to these islands.' D. /

4. CANIS AZARAE

PLATE VII

Canis Azarae, *Pr. Maximilian, Beiträge zur Naturgeschichte Braziliens*, vol. ii, p. 338.

Agouarachay, *Azara, Essais sur l'histoire naturelle des Quadrupèdes de la Province du Paraguay*, tom. i, p. 317.

C. *suprà albo nigroque variegatus; lateribus cinerescentibus; capite, auribus externè, artubusque, cinereo-cinnamominis; mento nigro; tibiis externis ad basin nigro lavatis; caudâ albescente, suprà nigro variegatâ, ad apicem nigrâ; spatio pone angulos oris, gutture, corporeque subtùs albescentibus; fasciis duabus griscescentibus in pectore plus minusve distinctis*

Description. Compared with the common fox (*Canis Vulpes*, Auct.), the present animal is rather smaller, and of a more slender form. Its limbs are a little longer in proportion; the ears are not so broad. The tail is not quite so bushy, neither is it so long; the fur is much longer, and of a harsher nature.

The predominant colours of the body are black and white; the limbs are of a fulvous hue externally. The hairs on the under part of the feet are dirty brown; the fore part of the anterior legs, and the feet, are of a buff colour; on the former, the hairs are more or less distinctly tipped with black, which produces a grizzled appearance. The inner side of the fore legs is of an uniform pale buff colour; the hinder part of these

PLATE VII

Canis Azarae

legs, the fore part of the posterior legs, and the inner side of the thighs, are white. On the outer side of the hinder legs, at some little distance above the heel, is a large blackish patch. The under parts of the body are of a dirty white hue, arising from the hairs being dusky or brownish at the base, and tipped with white, as on the fore part of the belly, or of a pale buff colour at the base, as towards the rump. The edge of the upper lip, the throat, neck, and chest, are white; a broad greyish band extends across the latter, and another of a paler hue crosses the lower part of the neck. The chin is black, and this colour is extended backwards around the angle of the mouth. The upper part of the head is of a pale yellow-brown colour, each hair being annulated with white near the apex. The ears are furnished with white hairs internally, and externally they are of a yellowish brown colour, tipped with black; at the base of the ears, and the portion of the neck on each side nearest to them, the / hairs are of an uniform buff colour. The hairs of the moustaches are long and stiff, and of a black colour. The hairs of the back, which are very long, are brown at the base, very pale towards the skin, and of a deep brown in the opposite direction; each hair is then white, and at the apex black. The tail is whitish, mottled with black; the apical portion is black, and there is a patch of the same colour towards the base on the upper side.

	Inches	*Lines*		*Inches*	*Lines*
Length from nose to root of tail	27	6	Length of tail (hair included)	14	6
from nose to base of ear	5	9	of ear	3	2
			Height of body at shoulders	14	0

Habitat, La Plata, Patagonia, and Chile.

The black and white portions of the hairs on the back produce in that part a mottled appearance, and in the specimen from which the above description is taken, these two colours are about equal in proportion. In another specimen now before me, the black colour predominates on the back. The fur in the younger animals of this species is not so long nor so harsh, and the upper parts are grizzled with black and white; that is to say, these two colours do not form patches of considerable extent as in the adults; the general colouring is also somewhat paler. The chin is brown-black or brown, instead of black, and the upper band, or that, which in the adult extends across the upper part of the neck, is interrupted in the middle; in fact, is only traceable on the *sides* of the neck.

Azara, in his description of the Agouarachay, says, the muzzle, as far back as the eyes, is blackish; whereas, in all the specimens examined by me, the muzzle is of the same colour as the other parts of the head, or *very* nearly so. In other respects his description agrees with the animal described by me, and *not* with the *Canis cinereo-argentatus*, which Desmarest and Lesson suppose to be the Agouarachay of Azara. In Fischer's *Synopsis Mammalium* the *Canis Azarae* is described as having the tip of the tail white; whereas it is black, not only in the five specimens which have come under my notice, but also in those in the collection of Prince Maximilian[8] (who was the original describer) and in the Parish Museum.

'This animal has a wide range; Prince Maximilian brought specimens from Brazil; and it is common in La Plata, Chile, the whole of Patagonia, even to the shores of the Strait of Magellan; and a fox, which lives on the small islands not far from Cape Horn, probably belongs to this species. This animal generally frequents desert places; I saw many in the valley of the Despoblado, a branch of / that of Copiapó, where there is no fresh water, and where, with the exception of some small rodents (the constant inhabitants of sterile regions), scarcely any other animal could exist. I saw also very many of these foxes wandering about by day (although Azara says they are nocturnal in Paraguay) on the plains of Santa Cruz, where various kinds of mice are abundant, and likewise around the Sierra Ventana. In the course of one day's ride in this latter neighbourhood (not far from Bahia Blanca, lat. 39°S), I should think I saw between thirty and forty. They generally were wandering at no great distance from their burrows; but, as they are not very swift animals, our dogs caught two. Azara states that in Paraguay this fox, which he calls the Agourá-chay, inhabits thick woods, and that it makes a great nest or pile of straw, to lie on; but that near Buenos Ayres it uses the holes of Bizcacha. Further southward, where the Bizcacha is not found, it certainly excavates its own burrow.[9] In Chile these foxes are very destructive to

[8] I am indebted to Mr Ogilby, who visted the Prince's collection, for a description from the specimens of *C. Azarae* therein preserved. In this description the tip of the tail is said to be black.

[9] Considering the great difference of climate and other conditions between the hot and wooded country of Paraguay, and the desolate plains of Patagonia, one is led to suspect that the *Canis Azarae* of La Plata and Patagonia, which wanders about by day, and inhabits burrows instead of heaps of straw, may turn out to be a different species from the Agourá-chay of Azara, which is nocturnal in its habits, and lives in thick coverts.

the vineyards, from the quantity of grapes they consume; so that boys are generally kept in the vintage season with bells and other means to frighten them away. Azara states, that in Paraguay they likewise eat fruit and sugar-cane. By the same authority it is said, that the Agourá-chay, when taken young, is easily domesticated.' D.

1. FELIS YAGOUAROUNDI

PLATE VIII

Felis Yagouaroundi, *Desmarest, Mammologie*, p. 230.

Yagouaroundi, *Azara, Essais sur l'histoire Naturelle des Quadrupèdes de la Province du Paraguay*, tom. i, p. 171.

Felis Darwinii, *Martin, Proceedings of the Zoological Society of London*, 1837, p. 3.

F. *vellere brevi, adpresso, purpurascenti-fusco; pilis flavescente annulatis; pedibus nigro lavatis; caudâ longissimâ; auribus parvulis.*

Description. The fur is rather harsh, short, and somewhat adpressed: the under fur is of a pale greyish brown colour; the hairs which constitute the chief clothing of the animal, are black, annulated with brownish yellow, or in some parts, yellow-white, each hair having about three or four rings. The black and pale colours are about equal in proportion, and their mixture produces / a deep brown tint, which is almost uniform throughout the body and limbs. On the head the yellowish colour predominates over the black, excepting on the tip of the muzzle, and thence back to the eye, where the hairs are of a brownish black colour. On the throat the hairs are brown. The underside of the tarsus is black, and on the outer side of the fore-foot there is a black mark which extends upwards on to the wrist. The tail is long and bushy; towards and on the base, the hairs are annulated with black and yellow, like those of the body; but beyond this they are of a more uniform colour, each hair being brown at the base, and gradually shaded into black towards the tip. The ears are small and rounded, and covered with hairs of the same colour as those on the head. The claws are of a large size, and white colour; the toes are united for a considerable portion of their length by the interdigital membrane.

PLATE VIII
Felis Yagouaroundi

	Inches	Lines		Inches	Lines
Length from nose to root of tail	25	0	Length of tail (hairs included)	19	0
from nose to base of ear	3	6	of ear	1	0
			Height of body at shoulders	12	0

Habitat, Rio de Janeiro, Brazil. (*May*)

'This cat was given me by an old Portuguese priest, who had hunted it down in a thick forest with a small pack of dogs, after a severe chase. It was killed near the Gavia mountain, at the distance of a few miles only from Rio de Janeiro, where it was considered uncommon.' D. Although small, compared with the Puma (*Felis concolor*, Auct.), this cat, in its slender lengthened body, small head, long tail, and stout limbs, decidedly evinces an affinity to that species. According to the dimensions of the Yagouaroundi given by Azara, Desmarest, and Temminck, it appears that the tail is considerably shorter in proportion in the specimens examined by those naturalists, than in the present individual, and the difference was such, as to induce Mr Martin to believe that the latter was a distinct species; he accordingly proposed for it the specific name of *Darwinii*. At the time that Mr Martin described the specimen alluded to, I was also inclined to believe it was a distinct species. I mention this because I am afraid my opinion had a slight share in influencing Mr Martin's determination. I have since seen many specimens, and upon comparing their dimensions, I find that the proportionate length of the tail varies more than is usual in other species of cats, and that the difference in the length in this member is not combined with any other distinguishing character. In colouring there is also a considerable variation, some specimens being almost black, and having the hairs but obscurely annulated with white; in others, the hairs are more distinctly annulated, and the head assumes a greyish hue. Others again, are brown, or black brown, / having the hairs annulated with yellow. The following are the dimensions of two specimens in the Paris museum, and those given by the authors above alluded to.[10]

	Paris M.		Paris M.		Desmarest.		Temminck.		Azara.	
	Inches	*Lines*	*Inches*	*Lines*	*Inches*	*Lines*	*Inches*	*Lines*	*Inches*	*Lines*
Length from nose to root of tail	30	6	28	0	23	0	30	0	36	9
of tail	24	0	17	0	13	9	22	0	13	0

[10] In measuring the species of Mammalia, I almost invariably, when wishing to give the length, measure from the tip of the nose *along the curve of the back* to the root of the tail. In the Ruminantia of course this plan is not desirable, but in other Mammals I have found it most convenient. If we take a Cat, for instance, and curve the body in whatever way we please, we find the length (taken in the way just mentioned) always the same. Whereas, if we take a straight line (as many naturalists do) the length will vary according to the position of the animal.

2. FELIS PAJEROS

PLATE IX

Chat Pampa, *Azara, Essais sur l'histoire Naturelle des Quadrupèdes du Paraguay. Traduct. Franç.*, tom. 1, p. 179.

Felis Pajeros, *Desmarest, Mammologie*, p. 231.

F. vellcre longissimo, flavescenti-griseo, fasciis flavescenti-fuscis indistinctè et sublongitudinalitèr notato; pedibus annulis latis nigris; abdomine maculis magnis nigris; mento albo; caudâ brevi; auribus mediocribus, ad apicem externum nigris.

Description. The Pampas cat is about equal in size to the common wild cat of Europe (*Felis Catus*, Linn.). It is however of a stouter form than that animal, the head is smaller, and the tail is shorter.

The most remarkable character in this species consists in the great length of the fur – the longer hairs on the back measuring upwards of three inches, and those on the hinder part of the back, are from four and a half, to four and three quarter inches in length. The general colour of the fur is pale yellow-grey. Numerous irregular yellow, or sometimes brown stripes run in an oblique manner from the back along the sides of the body. On each side of the face there are two stripes of a yellowish or cinnamon colour: these stripes commence near the eye, extend backwards and downwards over the cheeks, on the hinder part of which they join and form a single line, which encircles the lower part of the throat. The tip of the muzzle and the chin are white, and there is a spot in front of the eye, and a line beneath the eye, of the same colour: the belly and the inner side and hinder part of / the fore-legs are also white. An irregular black line runs across the lower part of the chest and extends over the base of the fore-legs externally, and above this line there are two other transverse dark markings on the chest, which are more or less defined. On the fore-legs there are three broad black bands, two of which encircle the leg, and on the posterior legs there are about five black bands externally, and some irregular dark spots internally. The feet are yellowish, and the underside of the tarsus is of a slightly deeper hue. On the belly there are numerous large irregular black spots. The ears are of moderate size, furnished internally with long white hairs; externally, the ears are of the same colour as the head, excepting at the apex where the hairs are black and form a slight tuft. The tail is short, somewhat bushy, and devoid of dark rings or spots – the hairs are in fact coloured as those of the back of the animal. On the upper part of the body each hair is brown at the base, then yellow, and at the apex, black. On the hinder part of the back the hairs are almost black at the base, and on the sides of the body each hair is grey at the base; there is then a considerable space of yellowish-white colour; towards the apex they are white, and at the apex black. The greater number of the hairs of the moustaches are white.

	Inches	*Lines*		*Inches*	*Lines*
Length from nose to root of tail	26	0	Length of tail (fur included)	11	0
from nose to base of ear	3	6	of ear	1	11
			Height of body at shoulders	13	0

Habitat, Santa Cruz, Patagonia (*April*), and Bahia Blanca (*August*).

The markings in this animal vary slightly in intensity; those on the body are generally indistinct, but the black rings on the legs are always very conspicuous.

'This animal takes its name from "paja", the Spanish word for straw, from its habit of frequenting reeds. It is common over the whole of the great plains, which compose the eastern side of the southern part of America. According to Azara, it extends northward as far as latitude 30°, and to the south, I have reason to believe, from the accounts I have received, that it is found near the Strait of Magellan, which would give it a range of nearly 1,400 miles, in a north and south line. One of my specimens was obtained, in 50° south, at Santa Cruz: it was met with in a valley, where a few thickets were growing. When disturbed, it did not run away, but drew itself up, and hissed. My other specimen was half-grown, and was killed in the end of August, at Bahia Blanca.' D. /

2. FELIS DOMESTICA

Felis domestica, *Brisson, Reg. Anim.*, p. 264.

I find in Mr Darwin's collection a cat, the colouring and proportions of which, convince me that its origin is from the domestic cat, as however

PLATE IX

Felis Pajeros

it was shot in a wild state far from any house, a description may, perhaps, prove useful. Its general colour is deep grey, and the body is adorned with numerous irregular narrow black bands; there is a broad black mark, formed of confluent spots, along the middle of the back, which commences a little behind the shoulders; a considerable space around the angles of the mouth, the chin, throat, central portion of the chest, fore-feet, toes of the hinder feet, and the posterior portion of the belly, are white; a black line extends backwards from the posterior angle of the eye, on to the cheeks; thence, across the throat, there are two lines: the space between the eye is chiefly occupied with white hairs: the tail is slender, and tapers towards the apex; the basal half is grey with black rings, and the apical half is black, excepting the extreme point, which is white: the tarsus is black beneath: the legs are of a deep grey colour, banded with black externally.

To the dimensions I will add those of a domestic cat which in colour and markings very closely resembles the animal above described. I may add that I have chosen a cat rather above the ordinary size for my comparison, yet it will be seen that the wild cat has the advantage in bulk.

	WILD CAT		DOMESTIC CAT	
	Inches	*Lines*	*Inches*	*Lines*
Length from nose to root of tail	22	0	19	0
of tail	12	3	11	6
of tarsus	5	1	4	7
of ear	1	11		
Height at shoulders	11	3		

Habitat, Maldonado, La Plata. (*May*)

'This animal was killed among some thickets on a rocky hill a few miles from Maldonado. It appeared, when dead, much larger and stronger than any domestic cat I ever saw, and it was described to me as having been exceedingly fierce. I mention this because M. Temminck supposes that the domesticated varieties of all animals are of larger size, than the wild stock from which they are descended.' D.[11] /

[11] I must refer the reader to my journal for some account of the habits of the jaguar and puma, which being well known animals, and the facts that I mention having little scientific interest, I have not thought it worth while to repeat them here.

GALLICTIS VITTATA

Gallictis vittata, *Bell, Zoological Journal*, vol. ii, pp. 551–2.
Gallictis vittata, *Bell, Proceedings of the Zoological Society*, for April, 1837, p. 89.
Gulo vittatus, *Desmarest, Mamm.*, p. 175.

'This animal is not uncommon at Maldonado, where it is called "*Huron*" or thief, from the ravages it commits on eggs and poultry. Shortly after being killed this specimen weighed 1 lb. 8 oz. (Imp. weight).' D.

1. LUTRA PLATENSIS

L. vellere nitido, adpresso, intensè fusco; corpore subtus pallidiore; gutture ad latera, et subtùs, pallidè fusco; mento rostrique apice sordidè flavescenti-albis; pedibus nigrescenti-fuscis; pilis caudae supernè brevioribus, adpressis, illis ad caudae latera longioribus et fimbriam efficientibus.

Description. This Otter is about equal in size to the common European species (*Lutra vulgaris*, Auct.): its fur is short, glossy, and adpressed; the under fur is tolerably abundant and of a silky nature. The general colouring of the ordinary fur is deep brown, and that of the under fur is very pale brown, deeper externally. The tint of the under parts of the body is paler than the upper, and may be described as brown, that of the throat, sides and under part of the neck, pale brown; and, on the tip of the muzzle and chin, dirty yellowish-white. The hairs of the moustaches are brownish-white; the ears are covered with short deep brown hairs, those towards the tip are paler. The hairs covering the feet above are short, and of a very deep brown colour. The tail is tolerably long, thick at the base, whence it gradually tapers to the apex. The hairs on the base of the tail resemble those of the body, but on the remaining portion, they are short, glossy, and very closely applied to the skin both on the upper and under surface, whereas those on the sides are longer, and form a kind of fringe. The tip of the muzzle and the soles of the feet are naked, with the exception of the hinder half of the tarsus.

	Inches	Lines		Inches	Lines
Length from nose to root of tail	28	0	Length of tail	18	0

Habitat, La Plata. (*July*) /

The La Plata Otter in its general colouring is of a somewhat deeper hue than the European species, the cheeks and throat instead of being nearly white are of a pale brown colour; the tail is longer in proportion, and tapers more gradually; the tip of the muzzle is naked, but the hairless portion is less than in that species, the boundary line between the naked part and the hair of the top of the muzzle forming almost a semicircle; the retiring extremities of this line touch the posterior angle of the nostril on each side, whereas in the common otter the boundary line of the hair of the muzzle is of a W-like form. The skull is figured in Pl. XXXV, figs 4, *a*, *b*, *c*, and *d*, and is compared with that of *L. Chilensis* in the next description.

'This specimen was killed by some fishermen a few miles from Maldonado, near the mouth of the estuary of the Plata, where the water is quite salt. I am not, however, by any means sure that it may not be a fresh-water species, which had wandered from its proper station; in the same manner as not unfrequently is the case with the *Hydrochaerus Capybara*. I am indebted to Mr Chaffers, the master of the *Beagle*, for having kindly presented me with this specimen.' D.

2. LUTRA CHILENSIS

Lutra Chilensis, *Bennett, Proceedings of the Committee of Science and Correspondence of the Zoological Society of London for 1832*, p. 1.

L. fusca; vellere mediocri, laxo et sub-extante; mento, gulâ, et faciei lateribus, pallidè fuscis; pedibus saturatè fuscis; corporis pilis ad apicem pallidè fuscis; caudâ mediocri; rostri apice calvo.

Description. This species scarcely equals a full grown European otter in size. It is of a brown colour throughout; the cheeks, chin, and throat, being slightly paler, and the feet of a deeper tint, than the other parts. The fur is moderately long, rather harsh to the touch, and semi-erect: the under fur is abundant, and of a soft and silk nature. The hairs of the ordinary fur are deep brown, but tipped with a very pale brown colour. The hairs of the tail, like those of the body, are harsh and semi-erect; towards the apex, those on the upper and under part are in a slight degree shorter than those at the sides, and lie closer to the skin; these differences, however, are not very apparent on the upper side, though distinct on the under. The feet are naked beneath, with the exception of the posterior half of the tarsus. The hair of the muzzle extends only down to the posterior angle of the / nostrils, where it terminates in a straight line, leaving the tip of the muzzle naked.

	Inches	*Lines*		*Inches*	*Lines*
Length from nose to root of tail	31	0	Length of tail	14	3

Habitat, Chonos Archipelago. (*January*)

The Chile Otter was originally described by Mr Bennett from a specimen presented to the Zoological Society by Mr Cuming, but as this specimen is a young animal, scarcely half-grown, it does not present some of the characters of the species in so marked a manner as the adult. I have, therefore, availed myself of an adult specimen in Mr Darwin's collection, to draw up the above description.

Compared with the Common Otter (*Lutra vulgaris*, Auct.) the most striking difference consists in the character of the fur: the hairs instead of being adpressed as in that species, are here semi-erect, and appear as if they had been clipped at the extremity. The fur is of a deeper colour, but has a slightly grizzled appearance, owing to the tip of each hair being of a much paler colour than the remaining part.

In the young animal described by Mr Bennett (which in weight was probably not more than one-third of that of the present animal), the hairs of the body are of an uniform deep brown colour; hence, if I am right in considering Mr Darwin's animal as the same species, it would appear that the grizzled character of the fur is dependent on age.

The semi-erect fur will also serve to distinguish the present species from the *Lutra Platensis*; the fur is likewise longer, the tail is shorter, and the feet are smaller in proportion. The most important distinctions, however, are furnished by the skulls; I will, therefore, compare them.

The skull of *L. Chilensis* compared with that of *L. Platensis* (Pl. XXXV, figs 4), when viewed from above, presents but little difference in general form; it is, however, smaller in all its proportions, and the zygomatic arch is a little less convex: the palate is proportionately shorter; the tympanic bullae are much smaller, less elevated, and wider apart, in which respect there is a greater approximation to the skull of *L. vulgaris* than to that of *L. Platensis*; but here, the tympanic bullae are larger than in *L. Chilensis*. Both in *L. Chilensis* and *Platensis*, the sub-orbital foramina are kidney-shaped, the emarginated portion being downwards, whilst in *L. vulgaris* they approach somewhat to a triangular figure, the apex being external. In *L. Chilensis*, however, this foramen is comparatively larger than in *L. Platensis*, and the outer portion of the foramen forms the / segment of a larger circle than the inner one, whilst in *L. Platensis* both portions are equal.

The principal difference in the dentition of the La Plata and the Chile otters, consists in the comparatively smaller size of the posterior molars, both of the upper and lower jaws, of the latter species. In the upper jaw, the 'carnassière' has its inner lobe, approaching somewhat to a triangular form, whereas in *L. Platensis* it is broader and almost semicircular. In the lower jaw, the last molar but one has the inner lobe much smaller than the middle outer lobe, whilst in *L. Platensis* these two lobes are of nearly equal size and elevation. Other points of dissimilarity will be perceived in the annexed table of admeasurements.

	L. Chilensis		*L. Platensis*	
	Inches	*Lines*	*Inches*	*Lines*
Whole length of skull	3	9¾	4	2½
Greatest width	2	6⅓	2	10⅓
Width of skull from the apex of one mastoid process to the opposite	2	3¼	2	8¼
Length of palate	1	6	1	10
Breadth of palate between the posterior molars		7¾		7¾
Length from last molar to posterior margin of palate		3⅓		5½
from base of canine to hinder part of last molar		11⅔	1	1⅓
of carnassière		5		5⅔
Width of carnassière		5		6¼
Length of last molar		2¾		3½
Width of last molar		4⅓		5⅔
Length of ramus of lower jaw	2	4½	2	8¾
from canine to hinder portion of last molar (lower jaw)	1	2¼	1	4¼
of last molar but one (lower jaw)		5¾		6⅔
Width of last molar but one (lower jaw)		2⅔		3½

'These animals are exceedingly common amongst the innumerable channels and bays, which form the Chonos Archipelago. They may generally be seen quietly swimming, with their heads just out of water, amidst the great entangled beds of kelp, which abound on this coast.

They burrow in the ground, within the forest, just above the rocky shore, and I was told, that they sometimes roam about the woods. This otter does not, by any means, live exclusively on fish. One was shot whilst running to its hole with a large volute-shell in its mouth; another (I believe the same species) was seen in Tierra del Fuego devouring a cuttle fish. But in the Chonos Archipelago, perhaps the chief food of this animal, as well as of the immense herds of great seals, and flocks of terns and cormorants, is a red coloured crab (belonging to the family Macrouri) of the size of a prawn, which swims near the surface in such dense bodies, that the water appears of a red colour. This specimen weighed nine pounds and a half.' D. /

FAMILY: DELPHINIDAE

DELPHINUS FITZROYI

PLATE X

D. suprà niger; capitis corporisque lateribus, corporeque subtùs niveis; caudâ, pedibus, labioque inferiore, nigris; fasciis latis duabus per latus utrumque obliquè excurrentibus, nigréscenti-cinereis, hujusque coloris fasciâ utrinque ab angulo oris ad pedem tendente

Description. Upper parts of the body black, under parts pure white, the two blended into each other by grey: extremity of snout, a ring round the eye, the edge of the under lip, and the tail fin, black; dorsal and pectoral fins dark grey; a broad grey mark extends from the angle of the mouth to the pectoral fin; above which, the white runs through the eye and is blended into grey over the eye; two broad deep-grey bands are extended in an oblique manner along each side of the body, running from the back downwards and backwards; iris of eye dark brown. Body anteriorly somewhat depressed, posteriorly compressed; head conical, arched above; the lower lip projecting beyond the upper; eye placed above and behind, but near the angle of the mouth; breathing vent situated in the same line as the eyes – supposing a circle to be taken round the head. Teeth slightly curved, and conical; in the upper jaw twenty-eight in number on each side, and in the lower, twenty-seven.

	Feet	Inches	Lines
Total length (measuring along the curve of back)	5	4	0
Length from tip of muzzle to vent	3	10	9
Length from tip of muzzle to dorsal fin	2	6	5
Length from tip of muzzle to pectoral	1	4	5
Length from tip of muzzle to eye	0	9	9
Length from tip of muzzle to breathing aperture (following curve of head)	0	10	7
Length from tip of muzzle to angle of mouth	0	7	9
Length of dorsal fin along the anterior margin	1	0	5
Height of dorsal fin along the anterior margin	0	6	4
Length of pectoral, along anterior margin	1	2	8
Width of tail	1	4	5
Girth of body before dorsal fin	3	0	6
Girth of body before pectoral fin	2	8	2
Girth of body before tail fin	0	7	8
Girth of head over the eyes	2	0	0

Habitat, coast of Patagonia, Lat. 42° 30′. (*April*) /

This species, which I have taken the liberty of naming after Captain FitzRoy, the Commander of the *Beagle*, approaches in some respects to the *Delphinus superciliosus* of the *Voyage de la* Coquille, but that animal does not possess the oblique dark-grey bands on the sides of the body; it likewise wants the grey mark which extends from the angle of the mouth to the pectoral fins. In the figure the under lip of the *D. superciliosus* is represented as almost white, whereas in the present species it is black: judging from the figures, there is likewise considerable difference in the form. The figure which illustrates this description agrees with the dimensions, which were carefully taken by Mr Darwin immediately after the animal was captured, and hence is correct.

'This porpoise, which was a female, was harpooned from the *Beagle* in the Bay of St Joseph, out of several, in a large troop, which were sporting round the ship. I am indebted to Captain FitzRoy for having made an excellent coloured drawing of it, when fresh killed, from which the accompanying lithograph has been taken.' D.

PLATE X
Delphinus FitzRoyi

FAMILY: CAMELIDAE

AUCHENIA LLAMA. *Desmarest*

Guanaco of the aborigines of Chile

'The Guanaco abounds over the whole of the temperate parts of South America, from the wooded islands of Tierra del Fuego, through Patagonia, the hilly parts of La Plata, Chile, even to the Cordillera of Peru. I saw several of these animals in Navarin Island, forty miles north of Cape Horn; the Guanaco, therefore, has, with the exception of a fox and mouse, inhabitants of the same island, the most southern range of all American quadrupeds. Although preferring an elevated site, it yields in this respect to its near relative the Vicuña. On the plains of Southern Patagonia, we saw them in greater numbers than in any other part. Generally they go in small herds, from half a dozen to thirty together; but on the banks of the Santa Cruz, we saw one herd, which must have contained at least five hundred. On the northern shores of the Strait of Magellan they are also very numerous. The Guanacoes are generally wild and extremely wary: Mr Stokes told me, that he one day in Patagonia saw through a glass a herd of these beasts, which evidently had been frightened, and were running away / at full speed, although their distance was so great that they could not be distinguished by the naked eye.

'The sportsman frequently receives the first intimation of their presence, by hearing from a long distance their peculiar shrill neighing note of alarm. If he then looks attentively, he will, perhaps, see the herd standing in a line on the side of some distant hill. On approaching, a few more squeals are given, and then off they set, at an apparently slow but really quick canter, along some narrow beaten track to a neighbouring hill. If, however, by chance he should abruptly meet a single animal, or several together, they will generally stand motionless, and intently gaze at him; then, perhaps, move on a few yards, turn round, and look again. What is the cause of this difference in their shyness? Do they mistake a man in the distance for their chief enemy the puma? Or does curiosity overcome their timidity? That they are curious is certain, for if a person lies on the ground, and plays strange antics, such as throwing up his feet in the air, they will almost always approach by degrees to reconnoitre him. It is an artifice that was repeatedly practised with success by the sportsman of the *Beagle*, and it had moreover the advantage of allowing several shots to be fired, which were all taken as parts of the performance. On the mountains of Tierra del Fuego, and in other places, I have more than once seen a Guanaco on being approached, not only neigh and squeal, but prance and leap about in the most ridiculous manner, apparently in defiance, as a challenge. These animals are very easily domesticated, and I have seen some in this state near the houses in northern Patagonia, although at large on their native plains. They are, when thus kept, very bold, and readily attack a man, by striking him from behind with both knees. It is asserted, that the motive for these attacks is jealousy on account of their females. The wild Guanacoes, however, have no idea of defence; and even a single dog will secure one of these large animals, till the huntsman can come up. In many of their habits they are like sheep in a flock. Thus when they see men approaching in several directions on horseback, they soon become bewildered, and know not which way to run. This circumstance greatly facilitates the Indian method of hunting, for they are thus easily driven to a central point, and are encompassed.

'The Guanacoes readily take to the water; several times at Port Valdes they were seen swimming from island to island. Byron, in his voyage, says he saw them drinking salt water. Some of our officers likewise saw a herd apparently drinking the briny fluid from a Salina near Cape Blanco; and in several parts of the country, if they do not drink salt water, I believe they drink none at all. In the middle of the day, they frequently roll in the dust, in saucer-shaped hollows. The males often fight together; one day two passed quite close to me, squealing and trying to bite each other; and several were shot with their / hides deeply scored. Herds appear sometimes to set out on exploring parties: at Bahia Blanca, where within thirty miles of the coast these animals are extremely scarce, I one day saw the tracks of thirty or forty, which had come in a direct line to a muddy salt water-creek. They then must have perceived, that they were approaching the sea, for they had wheeled with the regularity of cavalry, and had returned back in as straight a line, as they had advanced. The Guanacoes have one singular habit, the motive of which is to me quite inexplicable, namely, that on successive days they drop their dung on one defined heap. I saw one of these heaps, which was eight feet in diameter, and necessarily was composed of a large quantity. Frezier remarks on this habit as common to the Guanaco as well as to the

Llama;[12] he says it is very useful to the Indians, who use the dung for fuel, and are thus saved the trouble of collecting it.

'The Guanacoes appear to have favourite spots for dying in. On the banks of the Santa Cruz, the ground was actually white with bones in certain circumscribed spaces, which generally were bushy and all near the river. On one such spot I counted between ten and twenty heads. I particularly examined the bones; they did not appear, as some scattered ones which I had seen, gnawed or broken as if dragged together by a beast of prey. The animals in most cases, must have crawled, before dying, beneath and amongst the bushes. Mr Bynoe informs me, that during the last voyage, he observed the same circumstances on the banks of the Rio Gallegos. I do not at all understand the reason of this; but I may add, that the Guanacoes which were wounded on the plains near the Santa Cruz invariably walked towards the river. This quadruped seems particularly liable to contain in its stomach bezoar stones. The Indians who trade at the Rio Negro, bring great numbers to sell as Remedios or quack medicines; and I saw one old man with a box quite full of them, large and small. D. /

FAMILY: CERVIDAE

CERVUS CAMPESTRIS

Cervus campestris, *F. Cuvier*, in *Dict. des Sc. Nat.*, VII, p. 484.

Cervus campestris, *Cuvier, Oss. Foss.*, IV, p. 51, Pl. 3, f. 46.[13]

Guazuti, *Azara, Natural History of the Quadrupeds of Paraguay.* W. P. Hunter's translation, vol. i, p. 135.

Guazuti, *Azara*, French translation, vol. i, p. 77.

Besides skins of this species of stag, I find, in Mr Darwin's collection, three pairs of horns, which, together with a pair belonging to one of the skins, constitute a sufficiently complete series to illustrate the different forms which these appendages assume, as the animal increases in size.

[12] D'Orbigny says (vol. ii, p. 69), that all the species of the genus have this habit.

[13] Figures 47 and 48 of M. Cuvier's work represent horns so unlike either of those brought over by Mr Darwin, that I cannot help suspecting they belong to some other species of stag.

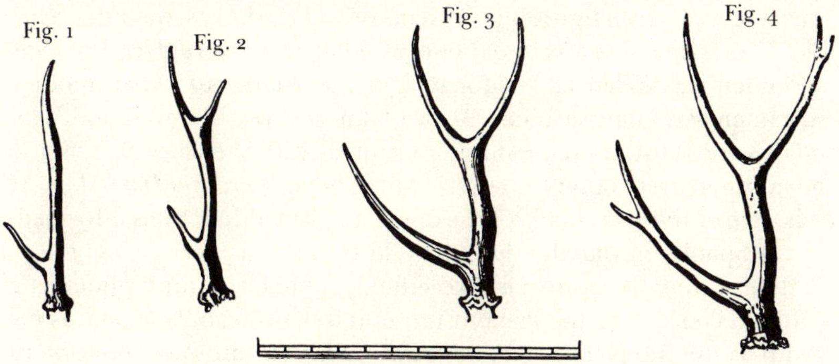

Scale of twelve inches

The above four sketches, which are all drawn to the same scale, will help to convey a clear idea of the forms, and relative proportions, of these horns.

The most simple horn (fig. 1) consists of a *beam*, eight and a half inches long, which is slightly arched outwards and considerably compressed about two and a half inches from the apex. At one inch from the base there is a small brow antler which projects forwards and upwards.

In the next horn (fig. 2), there is the same small brow antler, but there is a single small *snag*, about equal in size to the brow antler, which is directed backwards / and upwards, and is situated at three and a quarter inches from the apex of the beam. The total length of the beam is eight inches, measured in a straight line.

The third pair of horns (fig. 3), which must have belonged to an animal considerably older than either of the preceding pairs, exhibits a large brow antler, in length exceeding half that of the beam: here the posterior snag is also large, and is directed backwards and upwards, whilst the apical portion of the beam is directed forward about as much as the snag is directed backwards. The total length of this horn is eleven and a half inches, measured in a straight line.

The last figure (No. 4) represents the horn of one of the specimens of which an entire skin was brought over. This horn differs only from the last in being slightly larger, and in having two additional small snags, one springing from the under side, and near the apex, of the brow antler, and the other springing from the hinder part, and near the apex of the great posterior snag.

'The Spaniards say they can distinguish how old a deer is by the number of the branches on the horns. They affirmed that the specimen, of which figure 4 represents one of the horns, was nine years old. It certainly was a very old one, as all its teeth were decayed. This specimen was killed at Maldonado, in the middle of June; another specimen was killed at Bahia Blanca (almost three hundred and sixty miles southward), in the month of October, with the hairy skin on the horns: there were others, however, whose horns were free from skin. At this time of the year, many of the does had just kidded. I was informed, by the Spaniards, that this deer sheds its horns every year.

'The *Cervus campestris* is exceedingly abundant throughout the countries bordering the Plata. It is found in Northern Patagonia as far south as the Rio Negro (Lat. 41°); but, further southward, none were seen by the officers employed in surveying the coast. It appears to prefer a hilly country; I saw very many small herds, containing from five to seven animals each, near the Sierra Ventana, and among the hills north of Maldonado. If a person, crawling close along the ground, slowly advances towards a herd, the deer frequently approach, out of curiosity, to reconnoitre him. I have by this means killed, from one spot, three out of the same herd. Although thus so tame and inquisitive, yet, when approached on horseback, they are exceedingly wary. In this country nobody goes on foot, and the deer knows man as its enemy, only when he is mounted and armed with the bolas. At Bahia Blanca, a recent establishment in Northern Patagonia, I was surprised to find how little the deer cared for the noise of a gun: one day, I fired ten times, from within eighty yards, at one animal, and it was much more startled at the ball cutting up the ground, than at the report.

'The most curious fact, with respect to this animal, is the overpoweringly / strong and offensive odour which proceeds from the buck. It is quite indescribable: several times, whilst skinning the specimen, which is now mounted at the Zoological Museum, I was almost overcome by nausea. I tied up the skin in a silk pocket-handkerchief, and so carried it home: this handkerchief, after being well washed, I continually used, and it was, of course, as repeatedly washed; yet every time, when first unfolded, for a space of one year and seven months, I distinctly perceived the odour. This appears an astonishing instance of the permanence of some matter, which in its nature, nevertheless, must be most subtile and volatile. Frequently, when passing at the distance of half a mile to leeward of a herd, I have perceived the whole air tainted with the effluvium. I believe the smell from the buck is most powerful at the period when its horns are perfect, or free from the hairy skin.

When in this state the meat is, of course, quite uneatable; but the Spaniards assert, that if buried for some time in fresh earth, the taint is removed. These deer generally weigh about sixty or seventy pounds.' D.

FAMILY: MURIDAE

1. MUS DECUMANUS

Mus decumanus, *Auctorum*

In the extensive collection of Rodent animals brought home by Mr Darwin, I find several specimens of the above named species, that is to say, animals which resemble the European specimens of *Mus Decumanus* in all those characters which are the least liable to variation in individuals of the same species, such as the proportions which the various parts of the animal bear to each other: they differ, however, somewhat in colouring.

Buenos Ayres, Maldonado, Valparaiso, East Falkland Island, and Keeling Island, are each, it appears, infested with the common European rat. I have now before me two specimens from East Falkland Island, and one specimen from each of the other localities, and among these I find none equal in size to the largest European specimens: as regards the colouring, the Buenos Ayres specimen differs only from the English specimens of *Mus Decumanus*, in having the upper parts of a richer and deeper hue, owing to the tips of the shorter hairs being of a deep yellow instead of pale yellow, and in having a rusty tint over the haunches. /

Mr Darwin found this variety 'common about houses in the country around Buenos Ayres'.

In the Maldonado variety, the shorter hairs of the upper parts of the body are of a rusty yellow colour at the apex, in other respects it resembles the British variety. The rusty yellow colour of the tips of the hairs produces a general reddish hue, which is the more conspicuous, when the animal is placed near an English specimen. 'Was caught in a house, at Maldonado. I saw a specimen of the common grey English, or Norway rat, lying dead in the streets, and it certainly had a very different appearance from these red rats. The latter, I saw crawling about the hedges in the interior provinces at Santa Fé, and likewise in the forest of the island of Chiloe. This latter fact, however, is a strong argument against its being aboriginal, since I did not find even one undoubted American species, out of the many which I collected, inhabiting both sides of the Cordillera.' D.

The specimen from Valparaiso very closely resembles that from Maldonado; it is, perhaps, a little less red. 'Common about the houses in the town of Valparaiso.'

The two specimens from East Falkland are of a brighter hue, and have less grey in their colouring, than in the European variety of the common rat. 'One of them was caught in a Bay, which is sometimes frequented by shipping, but which is distant thirty or forty miles from any habitation. These rats have spread, not only over the whole of East and West Falkland, but even on some of the outlying islets. When the cold, wet, and gloomy nature of the climate is considered, it is surprising that these animals should be able to find food to live on.' D.

The general hue of the Keeling Island specimen, is deep brown, the longer hairs of the upper parts of the body being, as usual, black; but the shorter hairs, instead of having the pale yellow tint which we observe in the European (or, rather, British), specimens of *Mus Decumanus*, are of a deep, rusty yellow. The most remarkable difference, however, consists in the colouring of the under parts being of a yellowish tint, and, towards the root of the tail, of a very distinct buff yellow: the feet are brownish.

'This rat is exceedingly numerous on some of the low coral islets forming the margin of the Lagoon of Keeling Island, in the Indian Ocean. The climate is dry and hot. The rats are known to have come in a vessel from the Mauritius, which was wrecked on one of the islets, which is now called Rat Island. They appeared stunted in their growth, and many of them were mangy. They are supposed to live chiefly on cocoa-nuts, and any animal matter the sea may chance to throw up. They have not any fresh-water; but the milk of the cocoa-nut would supply its place.' D. /

The principal dimensions of the above animals are as follows:

	Specimen from Buenos Ayres		Maldonado		Valparaiso		East Falkland		East Falkland		Keeling Island	
	Inches	*Lines*	*Inches*	*Lines*	*Inches*	*Lines*	*Inches*	*Lines*	*Inches*	*Lines*	*Inches*	*Lines*
Length from nose to root of tail	9	9	9	3	8	6	8	9	9	0	8	3
of tail	Imperfect		6	0	6	6	Imperfect		6	0	6	6
of tarsus	1	7	1	7	1	7	1	7	1	7	1	7

Upon comparing the skull of the Valparaiso variety with that of a British specimen of *Mus decumanus*, I could perceive no difference. A skull from West Falkland did not differ, neither did the dentition of the Keeling Island specimen above noticed. A perfect specimen of this last I have not had an opportunity of examining.

2. MUS (DECUMANUS *var.*?) MAURUS

Mus maurus, *Waterh.* in *Proceedings of the Zoological Society of London*, for February, 1837, p. 20.

M. pilis suprà purpurescenti-nigris; subtùs plumbeis; auribus parvulis, pallidè fuscis: caudá corpus ferè aequante

Description. The character of the fur of this animal nearly resembles that of *Mus decumanus*; it is, however, of a harsher nature: the general colour of the upper parts and sides of the body is purple-black, arising from the longest hairs being of this colour, and likewise the tips of those which are next in length; the latter, however, excepting at the tip, are white, and this white is not entirely hidden, even when the hairs are in their ordinary position: on the head the hairs assume a brownish hue, and are tolerably uniform: the limbs, and under parts of the body, are of a deep grey colour, with a faint purple-brown wash: the under fur is grey: the ears are small, of a brown-white, or very pale brown colour, and furnished with minute brown hairs: the small, scattered, bristly hairs of the tail are of an uniform brownish-black colour. The hairs of the moustaches are black at the base, and greyish at the apex.

	Inches	*Lines*		*Inches*	*Lines*
Length from nose to root of tail	11	3	Length of tarsus	1	8
of tail	7	6	of ear	0	6¼
			from nose to ear	2	2

Habitat, Maldonado, La Plata. (*June*) /

This rat is very closely allied to *Mus decumanus*, and I think may possibly prove an extraordinary local variety of that animal. Having but one skin, and no skull, I am unable to satisfy myself on this point. Its size, as will be seen by the admeasurements, exceeds that of the common rat, or, rather, it exceeds ordinary specimens of that animal, for I have seen *some* which were equal to it.

'It was killed near Maldonado, where it frequented holes in the sand hillocks near the shore. It is likewise found on the island of Guritti. If ships are ever infested with these monstrous rats, the above-mentioned localities are very likely places to have received colonies by such means. An old male weighed fifteen ounces and three quarters. The ears of this rat, when alive, were of a pale colour, which made a singular contrast with the black fur of its body.' D.

3. MUS JACOBIAE

Mus decumanoïdes,[14] *Waterh.* in *Catalogue of the Mammalia preserved in the Museum of the Zoological Society of London.*

M. suprà fuscus, griseo-lavatus, subtùs albus: pedum pilis sordidè albis; caudâ corpore cum cupite paulò longiore; auribus mediocribus: pilis perlongis in dorso crebrè inter caeteros commixtis

Description. The general tint of the upper parts of this rat, is greyish-brown (very nearly resembling that of *Mus decumanus*); the longest hairs, which on the hinder portion of the back are one inch and a half in length, are black; the ordinary hairs are black at the apex, there is then, on each hair, a considerable space occupied by pale yellow, and the remaining, or basal portion, is greyish white; the under fur is grey: the hairs of the chin, throat, and under parts of the body, are white, and without any grey colour at the roots: the feet are covered with dirty greyish hairs: the tail, which is slender, is very sparingly furnished with minute black hairs, both above and beneath: the ears are of moderate size, of a brownish flesh colour, and, to the naked eye, appear to be destitute of hair. The hairs of the moustaches are most of them black at the base, and greyish at the apex. /

[14] The MS. name of *M. decumanoïdes*, which I had applied to this animal, has been changed, in consequence of my having seen a different species, with the same name attached, in the museum of the India House.

	Inches	*Lines*		*Inches*	*Lines*
Length from nose to root of tail	7	6	Length of tarsus	1	4 1/1
of tail	7	6	of ear	0	7½
			from nose to ear	1	7½

Habitat, James Island, Galapagos Archipelago, Pacific Ocean. (*October*)

This species is scarcely equal in size to a full grown common black rat (*Mus Rattus*), the head is rather shorter in proportion, the tarsi are smaller, and the tail is longer. In the character of the fur, and length of the hairs, it *very* closely resembles that species: the ears are larger than in *M. decumanus*, and about equal to those of *M. Rattus*. In having the hairs of the under parts of the body of an uniform colour (i.e. not grey at the base), it resembles the *Mus Tectorum* of Savi; but the large size of that animal, the greater length of the fur, and its colouring, all serve to distinguish it from the present species, which I may here observe, is truly an old world form, and very distinct from another species, also from the Galapagos, which is hereafter described.

'It is very common in James Island, but is not found on all the islands, if on any other in the Archipelago. Although its appearance is so like that of the common rat, yet its habits appear to be rather different: it is less carnivorous, and does not appear to be so strongly attached to the habitations of man. This island was frequented, about one hundred and fifty years since, by the vessels belonging to the Bucaniers; so that the common rat might easily have been transported here. And if a very peculiar climate, a volcanic soil, and strange food, can together produce a race, or strongly marked variety, there is every probability of such change having taken place in this case.' D.

4. MUS (RATTUS *var.*?) INSULARIS

M. suprà grisescenti, colore subtùs dilutiore; tarsis purpureo-nigris: caudâ corpus cum capite aequante: auribus mediocribus: vellere molli

Description. No. 1. The general colour of this animal is what might be termed black, there is, however, an obscure purple-brown hue on the upper parts of the body, and the sides and under parts have a greyish tint, the hairs covering the feet above are of an uniform deep purple-brown, almost black. All the hairs of the body are grey at the base: the hairs of the moustaches are long and numerous, and of a black colour, having one or two white hairs intermixed: the ears are of moderate size, and very sparingly furnished with / minute dark hairs: the tail is long and slender, and has small, scattered, bristly hairs, of a brown-black colour.

	Inches	*Lines*		*Inches*	*Lines*
Length from nose to root of			Length of tarsus	1	3½
tail	7	0	of ear	0	7
of tail	6	6	from nose to ear	1	6

No. 2. Hairs along the centre of the back chiefly black, and but obscurely annulated, near the apex, with deep yellow: towards the sides of the body, and over the haunches, the hairs are more distinctly annulated, and on the sides of the body they are of a pale yellow at the apex: on the under parts the hairs are grey, tipped with dirty yellowish white: the feet are of the same deep purple-brown hue as in the specimen first described.

Habitat, Ascension Island, Atlantic Ocean. (*July*)

These two animals not only differ in the colour of the fur, one being of a grizzled brownish colour, and the other black, but there is a considerable difference in the texture of the fur. In the black specimen, the fur is very soft and glossy, and the long hairs, which are abundant, are very slender. In the brown specimen, the fur is of a harsher nature, the long hairs are not so abundant, but longer, and less slender. On the other hand, they agree in size, dentition, the length of the head, tarsus, and ears, and differ but in a trifling degree (about three lines), in the length of the tail.

Upon comparing the Ascension Island specimens with *M. Rattus*, I find that, although in size they are about one third less, yet the teeth precisely agree, not only in form, but in size. The relative proportions of the head, ears, and tarsi, also agree. Besides the general colouring of the fur, they both differ in having the hairs of the feet uniformly purple-black, those in *Mus Rattus* being much paler, and even whitish, on the toes. In the character of the fur, there is much difference. The long silky hairs, which are so conspicuous in *Mus Rattus*, are replaced, in the black specimen, by hairs which are scarcely to be distinguished from the ordinary fur; and in the other specimen, although rather longer and more distinct, they are short, compared with those of the black rat.

'The specimen which has a black, and glossy fur, frequents the short coarse grass near the summit of the island, where the common mouse likewise occurs. It is often seen running about by day, and was found in numbers, when the island was first colonized by the English, a few years since. The other, and browner coloured variety, lives in the out-houses near the sea-beach, and feeds / chiefly on the offal of the turtles, slaughtered for the daily food of the inhabitants. If the settlement were destroyed, I feel no doubt that this latter variety would be compelled to migrate from the coast. Did it originally descend from the summit? and, in the case just supposed, would it retreat there? and, if so, would its black colour return? It must, however, be observed, that the two localities are separated from each other by a space, some miles in width, of bare lava and ashes. Does the summit of Ascension, an island so immensely remote from any continent, and the summit itself surrounded by a broad fringe of desert volcanic soil possess a small quadruped, peculiar to itself? Or, more probably, has this new species been brought, by some ship, from some unknown quarter of the world? Or, I am again tempted to ask, as I did in the case of the Galapagos rat, has the common English species been changed, by its new habitation, into a strongly marked variety?' D.

Mr Darwin seems to have foreseen the difficult problem which these two rats have furnished, and although I have spent much time in studying the Muridae, I must confess I have been exceedingly puzzled by the animals in question. It appears as if the brown, and black rats (*M. decumanus*, and *M. Rattus*), and likewise the common mouse (*M. Musculus*),[15] all of which follow man in his peregrinations, and which, to a certain degree, are dependent upon man, and may therefore be termed semi-domestic animals; like *really* domestic animals, are subject to a greater degree of variation than those species which hold themselves aloof from him.

Upon the whole then I have determined to describe the two Ascension Island specimens as one species, and as varieties of the *Mus Rattus*, but with a mark of doubt, since I do not possess sufficient materials for a rigorous examination, having, in fact, but one skin of each variety, and neither skull nor skeleton. I have also applied the name of *insularis*, to designate this variety or species, whichever it may

[15] The great Bandicoot rat of India (*Mus giganteus*, of Hardwicke), ought, perhaps, to be added to the species above enumerated; and I strongly suspect several catalogued species will prove but varieties of this animal.

be, for, supposing it be not a distinct species, it is so marked a variety, that a name for it is desirable. /

5. MUS MUSCULUS

Mus Musculus, *Auctorum*

Of this species, there are six specimens in Mr Darwin's collection; two were found 'living in the short grass, near the summit of the Island of Ascension, where the climate is temperate.' D. Two others were procured 'on a small, stony, and arid island, near Porto Praya, the capital of St Jago, in the Cape de Verde Islands – climate very hot and dry. Excepting during the rainy season, which is of short duration, these little animals can never taste fresh water, nor does the island afford any succulent plant.' D. A specimen was also procured 'on a grassy cliff, on East Falkland Island, at the distance of a mile from any habitation. It is singular that so delicate an animal should be able to subsist under the cold, and extremely humid climate, of the Falkland Islands, and on its unproductive soil.' D. These specimens are all of them rather less than full grown individuals of the same species procured in England; in other respects, they do not differ.

The sixth specimen, which is from Maldonado, is considerably less than British specimens of the common mouse, and is of a richer and brighter colour, the head is smaller, the muzzle shorter in proportion, whilst the tarsi are even longer than in a large specimen of *M. Musculus*. These points of dissimilarity induced me to believe it was a distinct species, and to apply to it the specific name of *brevirostris*.[16] Upon re-examination, with the advantage of more experience, and consequently a better knowledge of the characters of these animals, I have changed my opinion. The teeth indicate that it is not an adult specimen, and agree perfectly with those of *M. Musculus*, both in form and size. 'Common in the houses of the town of Maldonado, and its habits are similar to those of *Mus Musculus*.' D. /

[16] See *Proceedings of the Zoological Society* for 14 February, 1837, p. 19.

6. MUS LONGICAUDATUS

PLATE XI

Mus longicaudatus, *Bennett, Proceedings of the Committee of Science and Correspondence of the Zoological Society of London* for January, 1832, p. 2.

M. pallidè flavescenti-fuscus; corpore subtùs albo, levitèr flavo lavato; pedibus albis; tarsis permagnis; caudâ perlongâ; auribus parvulis

Description. Fur long and soft; general colour pale yellow-brown, the hairs of the ordinary fur being fulvous near the apex, and the longer hairs brown. On the sides of the body, cheeks, and external side of the limbs, the fulvous hue prevails. The inner side of the limbs and the under parts of the body are white, but have an indistinct yellowish hue. All the hairs of the body are of a deep grey colour at the base. The ears are small, well clothed with hairs; those on the inner side are chiefly yellow; externally, on the fore part they are brown, and posteriorly whitish. The feet are of a flesh-colour, and furnished above with white hairs; the tarsi are but sparingly provided with minute hairs on the upper side, and are naked beneath: they are of unusually large size. The fore feet are of moderate[17] size, and furnished with a very large carpal tubercle. The tail is very nearly double the length of the body, if the latter be measured in a straight line; it is of a brownish flesh-colour above, paler beneath, and sparingly furnished with minute bristly hairs; those on the upper surface being brown, and on the under side white. The hairs of the moustaches are long, of a black colour, and greyish at the apex.

	Inches	*Lines*		*Inches*	*Lines*
Length from nose to root of tail	3	9	Length of tarsus (claws included)	1	1
of tail	5	3	of ear	0	4
from nose to ear	0	10½			

Habitat, Chile. /

[17] As I shall have occasion to use the terms *moderate, long, short, large,* etc. it may be well to state that I take the common mouse (*Mus Musculus*), as my standard of comparison. The ears, feet, tail, length of the fur, general proportions, etc. are in that animal what I term moderate.

The most conspicuous characters of the present species consist in the immense length of the tail, and the great size of the hinder feet.[18] It is about equal in size to *Mus Musculus*; its form, however, is somewhat stouter; in colour it is much paler and brighter. The head is larger in proportion; the ears are smaller, and more densely clothed with hair; the fore feet are rather larger, and the fleshy tubercle on the under side of the wrist is also larger. The thumb nail is flattened, and rounded at the tip, as in *Mus Musculus*, but is longer, and more distinct than in that animal.

The skull of *M. longicaudatus* (Pl. XXXIV, Fig. 1), is considerably larger than that of the common mouse, but in form scarcely differs from it; its upper surface is rather more convex, and the interparietal bone proportionately less. The length of the skull is 1 inch; breadth, 6½ lines; distance between the fore part of the incisor, and the first molar of the upper jaw, 3½ lines. The dentition is figured in Pl. XXXIV, Figs 1 *b* and 1 *c*.

The above account is drawn up from the same specimen as that from which Mr Bennett took his description, and which was brought from Chile by Mr Cuming, who states that the animal in question lives in trees, and constructs its nest with grass.

In Mr Darwin's collection, I find an animal which agrees in all the more important characters with the one above described, but differs in being of a deeper colour (approaching more nearly, in this respect, to the common mouse), and in having the tail a trifle shorter. The skull is about ¾ of a line shorter, but its proportions agree precisely: the proportions of the feet, and the general form of the animal, also agree. This specimen is likewise from Chile (Lat 37° 40′), and, according to Mr Darwin, 'overran the wooded country south of Concepcion, in swarms of infinite numbers. Captain FitzRoy, on his return from visiting the wreck of H.M.S. *Challenger*, had the kindness to bring me this specimen. So destructive was this little animal, that it even gnawed through the paper of the cartridges belonging to the people who were wrecked.' D. /

[18] A long tarsus is generally accompanied by a proportionately long tail. I presume that those Mice which have long tarsi are in the habit of making great leaps, and that in these leaps, the tail serves to steady and balance the body.

PLATE XI

Mus longicaudatus *Mus gracilipes*

MUS ELEGANS

PLATE XII

Mus elegans, *Waterh., Proceedings of the Zoological Society of London* for February, 1837, p. 19.

Eligmodontia typus, *F. Cuvier, Annales des Sciences Naturelles* for March, 1837, tom. 7, p. 169, Pl. 5.

M. suprà flavus, vellere pilis fuscescentibus adsperso, his ad latera, et prope oculos rarioribus; pilis pone aurem utramque, labiis, corpore subtùs, pedibusque niveis; auribus magnis; caudâ capite corporeque paulo longiore; tarsis longis subtùs pilis obsitis.

Description. Fur very long and soft; general colour of the upper parts of the body pale brownish yellow; the lower portion of the cheeks, and the under parts of the body pure white: the hairs of the ordinary fur of the back are grey at the base, pale ochre near the apex, and brown at the apex; the longer hairs are brownish. On the sides of the body where the longer hairs are less numerous, the pale ochre colour prevails; the hairs on this part as on the back are deep grey at the base, but at a short distance from the apex they are white; nearer the tip shaded into yellow, and at the tip brownish: the limbs externally are of a pale yellow colour. The hairs of the throat and chest are pure white to the root, those on the belly are obscurely tinted with grey at the root. The feet are of a pale flesh-colour, and furnished with white hairs; the fore feet are of moderate size; the thumb nail is small and rounded, and the carpal tubercle is covered with hairs; the tarsi are long, and the white hairs extend over the whole of the under parts; the under side of the toes, however, are but sparingly furnished. There appears to be but one large tubercle on the under side of the tarsus, and this, which is situated near the base of the toes, is thickly covered with silvery-white hairs. The tail is long, pale brown above, and pale flesh-colour beneath; above, it is furnished with minute brown hairs, and on the under side with white hairs. The ears are rather large, of a pale flesh colour, tolerably well clothed with hairs, which are of a pale yellow colour on the inner side, and white on the outer side – excepting on the fore part, where they are brown. A small tuft of white hairs springs from the base of the ear posteriorly. The hairs of the moustaches are moderate; black at the base, and greyish at the apex. /

PLATE XII

Mus bimaculatus *Mus elegans*

	Inches	Lines		Inches	Lines
Length from nose to root of tail	3	7	Length from nose to ear	1	0
of tail	3	9	of tarsus	0	10
			of ear	0	6

Habitat, Bahia Blanca. (*September*)

Upon comparing the skull (Pl. XXXIV, fig. 2, *a*) of *M. elegans* with that of *M. Musculus*, the most evident points of distinction consist in the greater proportionate length of the nasal and frontal bones, and the slenderness of the zygomatic arch in the former animal. Length of skull 11 lines, width 6 lines, distance between front molar and outer side of incisors of upper jaw 3⅜ lines, length of nasal bones 4⅜ lines.

The dentition is figured in Pl. XXXIV, figs 2 *b*, and 2 *c*.

'Whilst bivouacking one night on shore, amongst some sand hillocks, this mouse, with its tail singed, leapt out of a bush which was placed on the fire. Its hind legs appeared long in proportion to the front, and it did not appear to be very active in endeavouring to make its escape.' D.

Mus elegans is about equal in size to *M. Musculus*; the head is larger in proportion than in the latter, the ears are slightly larger, the tail is longer, and so are the tarsi. The large ears, long tail, and comparatively large size of the feet, combined with the greater size of the animal itself, will render it easy to distinguish this species from *M. gracilipes* and *M. bimaculatus*. From the last mentioned animal it moreover differs in having the head larger in proportion, the fur longer, and the colouring of the upper parts of the body somewhat darker. The white fur is almost confined to the under parts of the body, and there is but a small tuft of white hairs behind the ears, whereas in *M. bimaculatus*, the white fur extends considerably on the sides of the body, the outer side of the limbs are white, and there is a large and conspicuous white spot behind each ear.

In *M. elegans* the whole sole of the tarsus and the carpal tubercles are covered with hair. In *Mus bimaculatus* the hinder *half* of the tarsus only is covered with hair, and in *M. gracilipes* both the hinder half is covered, and there are some scattered hairs extending almost to the two tubercles, which are situated at the base of the longer toes.

The genus *Eligmodontia* of M. F. Cuvier, founded upon a species of mouse from Buenos Ayres, possesses nearly the same characters as the subgenus *Calomys*, established by me in the *Proceedings of the Zoological Society* for February, 1837, and which included the animal above described, and two other species (*M. bimaculatus* and *M. gracilipes*). M

Cuvier's genus is distinguished by there being only one large tubercle on the under side of the tarsus, and in having the carpal pad covered with hair as well as the pad of the tarsus. In / these characters our present animal agrees, as it does also in size and in the relative proportions of the tail and tarsus, circumstances which induce me to believe they are identical.

In *M. bimaculatus* and *M. gracilipes* there are six naked tubercles on the under side of the tarsus, and the carpal pad is also naked. In having, however, the tarsus hairy beneath,[19] in dentition and in colouring, they agree so closely with *M. elegans* that I think they cannot be separated generically.

MUS BIMACULATUS

PLATE XII

Mus bimaculatus, *Waterh., Proceedings of the Zoological Society of London* for February, 1837, p. 18.

M. vellere pallidè ochraceo, pilis nigricantibus adsperso, his ad latera rarioribus; rostri lateribus, notâ magnâ pone aurem utramque, artubus, corporeque subtùs niveis; auribus mediocribus; caudâ, quoad longitudinem, corpus fere aequante; tarsis ad calcem pilis argenteo-candidis obsitis

Description. Upper parts of the body of a very pale ochre colour, the longer hairs, however, are black, and at the apex greyish, and where they are numerous, as on the back and upper surface of the head, they give greater depth to the colouring; the cheeks and sides of the body are of an almost uniform pale, but bright yellow; the sides of the muzzle, the lower half of the cheeks, the lower portion also of the sides of the body, and the whole of the under parts, are pure white – each hair being uniform in colour to the root, and not, as is usually the case, *grey* at the root. There is likewise a large patch of pure white hairs behind each ear. The feet and tail are of a pale flesh-colour, and furnished with white hairs, with the exception of those on the upper surface of the latter, which are pale brown. The ears are also pale flesh-colour, clothed internally with yellow hairs; externally on the fore part, the hairs are brownish, and on the hinder part, white – they

[19] In *Mus leocopus* of North America the tarsus is hairy beneath, and in the character of the teeth this animal also agrees with the species above mentioned.

are rather large, and so are the feet. The tail is about equal to the body in length. The hairs of the moustaches are numerous and slender, and most of them are black at the base, and grey at the apex. The hinder half of the tarsus / beneath is covered with minute silvery-white hairs; beside the ordinary tubercles, the anterior portion of the sole of the foot and the base of the toes beneath, are crowded with small rounded warts, which are much more numerous and conspicuous than in the common mouse.

	Inches	*Lines*		*Inches*	*Lines*
Length from nose to root of tail	3	1	Length from nose to base of ear	0	8 3/1
of tail	1	11	of tarsus (claws included)	0	8
from nose to eye	0	4½	of ear	0	4½

Habitat, Maldonado, La Plata. (*June*)

The skull of this animal, is rather shorter and broader than that of *Mus Musculus*, the upper surface is more arched, the zygomatic arch is much more slender, and the nasal bones are rather broader. In the convexity of the upper surface, and the slenderness of the zygomatic arch, this skull very nearly resembles that of *M. gracilipes*; this latter, however, has the zygomatic arch more convex, projecting more suddenly on the anterior part, and the interparietal bone smaller. Length of skull 10 lines, width 5½, length of nasal bones 4 lines, distance between the outer side of the incisors, of the upper jaw, and the first molar 2⅞ lines. See Pl. XXXIV, fig. 3 *a*.

The dentition is figured in Pl. XXXIV, figs 3 *b* and *c*.

This mouse is rather less than *M. Musculus*, the tail is much shorter in proportion, the fur is longer and softer, and the ears are more distinctly clothed with hair.

The pale and delicate yellow colour of the upper parts of the body, and the pure white of the under parts, renders the present species conspicuous amongst its congeners. I may further remark that the white colour which in the Muridae (when it occurs) is usually confined to the under part of the body, or extends but slightly on the sides, is in the present animal extended considerably on the sides of the body, and occupies an equal portion with the yellow of the upper parts. The name *bimaculatus* is applied to this animal on account of the two conspicuous white patches, which are situated behind the ears.

In affinity as well as in appearance it most nearly approaches to *Mus gracilipes* and *M. elegans*; with no other species of the genus Mus, here described, can it be confounded, since these only have the tarsus hairy beneath.

The principal points of distinction between the present animal and *Mus elegans*, are noticed in the account of that species.

'This mouse, when alive, had a very elegant appearance. A countryman, who brought it me, found six of them living together in one burrow.' D. /

MUS GRACILIPES

PLATE XI

Mus gracilipes, *Waterh.*, *Proceedings of the Zoological Society of London* for February, 1837, p. 19.

M. suprà flavo-lavatus; pilis pone aurem utramque, labiis, corporeque subtùs, albis; pedibus parvulis, gracilibus, carneis, suprà et ad calcem pilis albis tectis; candâ gracili, pilis albis instructâ, quoad longitudinem corpus ferè aequante; auribus mediocribus; vellere mediocri et molli, pilis omnibus ad basin plumbeis

Description. General colour very pale yellowish brown, a tint produced by the admixture of black and pale fawn colour; the hairs of the ordinary fur being of the latter tint near the apex, and dusky at the apex, whilst the longer hairs are black. The feet, tail, under parts of the body and the sides of the muzzle, are pure white. All the hairs of the body (which are soft, and of moderate length), are deep grey at the base. The ears are of moderate size, well clothed with hairs, of which those on the inner side are yellowish, and those on the outer, are brown on the anterior part, and white on the posterior. A small tuft of white hairs springs from the neck immediately behind the ears; this tuft is hidden when the ears are folded back. The tail is slender and short (being not quite equal to the body in length), of a pale flesh-colour, and sparingly furnished with minute white hairs. The feet are very small and slender, and the naked parts are of a pale flesh-colour. The sole of the foot is covered with hairs; the toes beneath, and the tubercles (which are as in *Mus Musculus*), however, are naked. The hairs of the moustaches are of moderate length, and of a blackish colour, some of them, however, are greyish white.

	Inches	Lines
Length from nose to root of tail	2	10
of tail	1	7
from nose to eye	0	4⅓
Length from nose to ear	0	8 1/1
of tarsus (claws included)	0	6½
of ear	0	4 1/1

Habitat, Bahia Blanca. (*September*)

This species slightly exceeds the harvest mouse (*Mus messorius*) in size, its ears are considerably larger in proportion, and the tail is shorter. Compared with the common mouse (*Mus Musculus*) it is smaller, the tail is more slender, and shorter, and the feet are likewise more slender and proportionately much smaller; the ears are more distinctly clothed with hairs. /

The principal points of distinction between this and the two preceding species are pointed out in the account of *M. elegans*.

Upon comparing the skull of *M. gracilipes* (Pl. XXXIV, fig. 4 *a*) with that of *Mus Musculus*, the most striking differences consist in its shorter and broader form, the upper surface being more arched, the inter-parietal bone has a relatively smaller antero-posterior diameter, the occipital region is more convex, and continued more gently and gradually into the upper region of the skull. The zygomatic arch, which is unusually slender, is more dilated (especially on the anterior part) thus giving a squareness to the general form. The nasal bones are not so much attenuated posteriorly. The length of the skull is 8⅞ lines, the greatest width is 5⅛ lines, and the distance between the outer side of the incisors and the front molar is 2¾ lines.

The dentition is figured in Pl. XXXIV, figs 4 *b* and 4 *c*.

'This specimen was given me by Mr Bynoe, the surgeon of the *Beagle*, who caught it amongst some long dry grass.' D.

MUS FLAVESCENS

PLATE XIII

Mus flavescens, *Waterh.*, *Proceedings of the Zoological Society of London*, for February, 1837, p. 19.

M. suprà colore cinnamomeo, lateribus capitis, corporisque, aequè ac pectore, auratis; gulâ abdomineque flavescenti-albis: pedibus sordidè albis: auribus mediocribus rotundatis, pilis flavis obsitis: caudâ, corpore, capiteque longiore, suprà fuscâ, subtùs sordidè albâ: tarsis longis

PLATE XIII

Mus flavescens *Mus arenicola*

Description. Fur long and moderately soft; general colour of the upper parts bright brownish yellow; on the sides of the head and body bright yellow; towards the rump of a deeper hue, and inclining to orange; under parts pale yellow, or yellow-white; chest yellow. The fur both of the upper and under parts of the body deep plumbeous at the base. Feet flesh colour, covered above with white hairs: tarsi long, naked beneath. Ears small, tolerably well clothed with hairs; those on the inner side yellow, but many of them blackish at the base; on the outer side, the hairs are blackish on the fore part and yellow on the hinder part. The hairs of the ordinary fur of the back are of a deep rich yellow colour at the tip, and the longer hairs are blackish. The tail is long, deep brown above and whitish beneath; the hairs of the / moustaches are rather short and slender, and of a brownish colour. Thumb nail small and rounded.

	Inches	*Lines*		*Inches*	*Lines*
Length from nose to root of tail	3	9	Length from nose to ear	1	0
of tail	4	1½	of tarsus	1	0½
			of ear	0	4½

Habitat, Maldonado, La Plata. (*June*)

This species is slightly larger than the common mouse; the head is rather larger in proportion; the ears are rather smaller and more distinctly clothed with hair; the tail and tarsi are much longer in proportion. Its bright yellow colouring and proportions distinguish it from any of the species described in this work. Of this animal I do not possess the skull, nor of the teeth do I possess more than the first and second molars of the upper jaw, and the second and last of the lower jaw. These are figured in Pl. XXXIV, figs 5 *a*, and 5 *b*.

MUS MAGELLANICUS

PLATE XIV

Mus Magellanicus, *Bennett, Proceedings of the Zoological Society of London* for December, 1835, p. 191.

M. suprà fuscus, subtùs cinerescenti-albus, pallidè flavo lavatus; auribus mediocribus pilis fuscis obsitis; caudâ corpus caputque aequante; tarsis longis, pilis sordidè albis obsitis

PLATE XIV

Mus brachiotis *Mus Magellanicus*

Description. Fur very long and moderately soft, general colour deep brown; the hairs of the ordinary fur are grey, tipped with yellowish brown; the longer hairs are black; the sides of the body are yellowish; the under parts are grey-white with a faint yellowish tint, each hair being grey tipped with yellowish white. The ears are rather small, well clothed with hairs; those on the inner side are blackish tipped with yellow, and on the outer side they are blackish on the fore part and dusky on the hinder part. The fore feet are of moderate size, the thumb nail is short and rounded; the tarsi are rather long; both fore and hinder feet are of a brownish colour, and covered above with dirty grey hairs. The tail rather exceeds the head and body in length; it is brown above and dirty white beneath. The hairs of the moustaches are numerous and long, of a brownish colour at the apex and black at the base. /

	Inches	*Lines*		*Inches*	*Lines*
Length from nose to root of tail	4	3	Length from nose to ear	1	0½
of tail	4	2	of tarsus	1	1
			of ear	0	5

Habitat, Port Famine, Strait of Magellan.

This mouse is larger than *Mus Musculus*; the tail is rather longer in proportion; the tarsi much longer; the ears are not quite so large in proportion to the head (which greatly exceeds that of *Mus Musculus* in size), and they are densely clothed with hair. The fur is longer. In colour, the animal here described is rather darker than the common mouse. I have one specimen however before me which *very nearly* agrees in this respect.

The dentition is figured in Pl. XXXIV, figs 6 *a* and 6 *b*.

From the attention which Mr Darwin bestowed upon the Muridae of the southern parts of South America, I presume his collection affords materials for a tolerably complete monograph of the species of that portion of the globe. The species above described, however, does not occur in Mr Darwin's collection, but is here introduced in order to make the work more complete, and that I might more clearly point out the distinctions which exist between it and other species here described, the account given by Mr Bennett in the *Proceedings* being very short.

MUS ARENICOLA

PLATE XIII

Mus arenicola, *Waterh.*, *Proceedings of the Zoological Society of London*, for February, 1837, p. 18.

M. suprà fuscus, subtùs cinerescenti-albus, pallidè flavo tinctus; auribus mediocribus rotundatis, pilis flavis fuscisque obsitis; caudâ quoad longitudinem corpus aequante; pedibus cinerescenti-albis: tarsis mediocribus

Description. Fur long, moderately soft; general colour deep brown; sides of the body with a very obscure yellowish hue; under parts dirty grey with a faint yellow tint. All the fur deep grey at the base; the hairs of the upper part of the body obscurely annulated with yellowish brown near the apex, and dusky at the apex; the longer hairs are black. Feet brownish, covered above with brown-white hairs; tarsi short. Tail short, blackish above, brown-white beneath. Ears small, well clothed with hairs; those on the / inner side are yellow at the apex and grey at the base; on the outer side they are of a brownish colour, and on the fore part blackish. The hairs of the moustaches are short and slender, and of a brownish colour. The head is large.

	Inches	*Lines*		*Inches*	*Lines*
Length from nose to root of tail	4	3	Length of tarsus (claws included)	0	10
of tail	2	9	of ear	0	4½
from nose to ear	1	0			

Habitat, Maldonado, La Plata. (*June*)

This species is rather larger than the common mouse; its head is proportionately larger, the ears are smaller, the tail considerably shorter, and the fur longer, and in colouring it is a little darker. In size and colour it resembles *M. Magellanicus*, but the shorter tail and tarsi, and the smaller size of the ears will serve to distinguish it.

The skull of *Mus arenicola*, Pl. XXXIV, fig. 7 *a*, is rather larger than that of *Mus Musculus*, the nasal portion is broader, the interparietal bone is much smaller, especially in antero-posterior extent; the zygomatic arches are more slender, and the incisive foramina are broader. The horizontal ramus of the lower jaw (Pl. XXXIV, fig. 7 *d*) is rather less curved, the coronoid process is more elongated, and the condyloid is narrower and also larger. The length of the skull is 11 lines and a half; the width is 6⅕ lines. The molars of the upper jaw are figured in Pl. XXXIV, fig. 7 *b* and those of the under jaw, fig. 7 *c*.

'This specimen was caught on the open grassy plain, by a trap baited with a piece of bird; it is, however, very abundant in the sand hillocks near the coast of the Plata.' D.

13. MUS BRACHIOTIS

PLATE XIV

Mus brachiotis, *Waterh.*, *Proceedings of the Zoological Society of London* for February, 1837, p. 17.

M. suprà obscurè fuscus, subtùs obscurè griseo tinctus; pedibus griseo-fuscis; auribus parvulis; caudâ quoad longitudinem, corpus ferè aequante: vellere longo et molli

Description. Fur soft, very long, and dense; ears very small; general colour brown: the hairs of the upper parts, and sides of the head and body are of / a deep grey at the base, black at the apex, and narrowly annulated with deep yellow near the apex; on the throat and belly they are of a paler grey at the base, and greyish white at the apex. The ears are well clothed with brown hairs both within and without, and are for the most part hidden by the long fur of the head. The hairs covering the upper side of the feet are of a palish ashy-brown colour, and the fleshy portion appears to have been brown. The tail is well clothed with hairs, so that the scales are scarcely visible; on the upper side of the tail the hairs are brownish-black, and on the under side, they are dirty white. The incisors are very slender; those of the upper jaw are of a very pale yellow colour, and those of the lower are white, or nearly so. The muzzle is slender, and pointed.

	Inches	*Lines*		*Inches*	*Lines*
Length from nose to root of tail	4	9	Length of tarsus (claws included)	0	11
of tail	2	8	of ear	0	3
from nose to base of ears	1	2			

Habitat, Chonos Archipelago. (*December*)

This mouse is considerably larger than *Mus Musculus*, and the great length and density of its fur, causes it to appear much stouter in its proportions; its colouring is darker, the tips of the hairs being much more narrowly annulated with yellow than in that species. The very small size of the ears will serve to distinguish the present animal from its congeners – *Mus longipilis*, *M. Renggeri*, *M. arenicola*, etc.

The molar teeth of the upper jaw are figured in Pl. XXXIV, fig. 8 *a*; and fig. 8 *b*, represents the middle and last molars of the lower jaw.

'Inhabited a very small island, covered with thick forest, in the central part of the Chonos Archipelago.' D.

A mouse obtained on the islets adjoining the east coast of Chiloe (where Mr Darwin says it was common) differs from the above in being a little smaller, the tail is rather longer, and the ears are a trifle larger. In the feet, claws, colouring and character of the fur it agrees, and likewise in the pale colour and slenderness of the incisors. Its dimensions are as follows:

	Inches	*Lines*
Length from nose to root of tail	4	0
of tail	3	0
Length of tarsus (claws included)	0	10
from nose to ear	0	10½
of ear	0	4

I have not the means of satisfying myself whether this be a distinct species or not; but I think it is not. /

'The nature of the country where this specimen was procured is nearly the same as in that part of the Chonos Archipelago, 150 miles to the south, where the first was obtained.' D.

14. MUS RENGGERI

PLATE XV – Fig. 1

Mus olivaceus, *Waterh.*, *Proceedings of the Zoological Society of London*, for February, 1838, p. 16.

M. corpore suprà subolivaceo, subtùs cinerescente; auribus mediocribus, rotundatis, pilis parvulis fuscescentibus obsitis; caudâ corpore breviore, pilosâ, suprà fuscâ subtùs albescente; pedibus pilis fuscescentibus tectis

Description. Fur moderate; ears moderate; tail shorter than the body; general colour grey washed with yellow; under parts greyish white. On the upper parts and sides of the head and body the hairs are grey, broadly annulated with yellow near the apex, and dusky at the apex; the mixture producing a yellowish grey tint, approaching somewhat towards olive: the hairs on the under parts of the body and throat are deep grey at the base, and white at the apex; the hairs of the feet are brownish white. The tail is tolerably well clothed with hairs; those on the upper surface are brown, and those on the under are dirty white. The ears are well clothed, both externally and internally, with hairs of the same colour as those on the upper parts of the body. The hairs of the moustaches are for the most part whitish, and black at the base. The upper incisors are pale yellow, and the lower incisors are yellowish white.

	Inches	*Lines*		*Inches*	*Lines*
Length from nose to the root of tail	5	1	Length of tarsus (claws included)	0	11
of tail	2	8	of ear	0	5
from nose to base of ears	1	2			

Habitat, Valparaiso (*August and September*), Coquimbo (*May*).

Subsequent to the description of this species, under the name of *M. olivaceus* in the Zoological Society's *Proceedings*, I have imagined that perhaps that name might mislead as regards the colouring of the animal; it certainly has a slight olive hue, but it is not very evident. I have therefore changed the name, and substituted that of the author of the *Naturgeschichte der Säugethiere von Paraguay*, etc. /

In the collection there are three specimens of the present species; in

PLATE XV

Mus Renggeri *Mus obscurus*

one of these the hairs of the upper part and sides of the body are annulated with yellowish white, instead of yellow; hence the general hue of these parts is nearly grey.

Mus Renggeri is larger than *Mus Musculus*, and much stouter in its proportions; the fur is shorter, much less dense, and less soft than in *Mus brachiotis*.

'It inhabits dry stony places, where only a few thickets grow.' D.

15. MUS OBSCURUS

PLATE XV – Fig. 2

Mus obscurus, *Waterh.*, *Proceedings of the Zoological Society of London* for February, 1837, p. 16.

M. suprà fusco-nigrescens, subtùs flavescens; pedibus obscurè fuscis; unguibus longiusculis; auribus mediocribus; caudâ corpore breviore, suprà nigrescente, subtùs sordidè albâ; vellere mediocri, molli

Description. Head large; ears moderate; tail shorter than the body; fur rather long and glossy; the general hue of that of the upper parts and sides of the head and body is blackish brown, and that of the under parts is dirty yellowish white. The hairs on the upper parts are of a deep lead colour at the base, black at the apex, and narrowly annulated with dark yellow near the apex; those of the throat and belly are lead colour at the base and yellowish at the tip; the chin is white: around the eye, and on the lower part of the cheeks a deep yellow tint prevails. The ears are well clothed with hairs both externally and internally, and these are for the most part of a deep brown colour, as are also the hairs which cover the feet. The tail is well clothed with hairs, those on the upper surface are black, and those on the under are dirty white. Both upper and lower incisors are yellow, but the lower are paler than the upper.

	Inches	*Lines*		*Inches*	*Lines*
Length from nose to root of tail	5	3	Length of tarsus (claws included)	0	11½
of tail	2	7	of ear	0	4
from nose to ears	1	2½			

Habitat, Maldonado, La Plata. (*June*)

The present species, like the foregoing, is much stouter than the common / mouse (*Mus Musculus*), its colour is much darker. In possessing a glossy fur it differs from most of its congeners; its head is also proportionately larger, and the incisors are much stronger.

The molars of the upper jaw are figured in Pl. XXXIV, fig. 9 *a* – and fig. 9 *b*, represents those of the under jaw.

'Very abundant in gardens and hedges, far from houses; and was easily caught in traps baited either with cheese or meat.' D.

16. MUS XANTHORHINUS

PLATE XVII – Fig. 1

Mus xanthorhinus, *Waterh.*, *Proceedings of the Zoological Society of London*, for January, 1837, p. 17.

M. suprà fuscus flavo lavatus; subtùs albus; rhinario flavo; auribus parvulis, intùs pilis flavis obsitis; mystacibus longis, canis, ad basin nigrescentibus: caudâ corpore breviore, suprà fuscâ, ad latera flavescente, subtùs sordidè albâ: pedibus anticis, tarsisque flavis, digitis albis: vellere longo, molli

Description. Fur moderately long and loose; ears rather small; tail shorter than the body; general colour grey washed with yellow, the yellow colour prevailing, especially on the sides of the body; muzzle, inner side of ears, and tarsus, of a rich yellow colour; toes, chin, throat, under parts of body, and rump, white; all the fur deep grey at the base; the hairs on the upper parts and sides of the body broadly annulated near the apex with rich yellow, and at the apex dusky; on the under parts of the body the hairs are broadly tipped with white. Tail rather sparingly furnished with hair, that on the upper surface brown, on the sides yellow, and on the under surface whitish. The hairs of the moustaches are white – some of them dusky at the base. The incisor teeth are rather slender, and of a pale yellow colour.

	Inches	*Lines*		*Inches*	*Lines*
Length from nose to root of tail	3	6	Length of tarsus (claws included)	0	9
of tail	1	7½	of ear	0	3¾
from nose to ear	0	10			

Habitat, Hardy Peninsula, Tierra del Fuego. (*February*)

The white, which is usually confined to the under parts of the body, in this / species extends slightly on the sides of the body, and the lower portion of the cheeks.

'This species was caught on the mountains, thickly covered with peat, of Hardy Peninsula, which forms the extreme southern point of Tierra del Fuego.' D.

17. MUS CANESCENS

Mus canescens, *Waterh., Proceedings of the Zoological Society of London* for February, 1837, p. 17.

M. suprà canescens, subtùs albus; oculis flavido cinctis; auribus parvulis, pilis pallidè flavis et plumbeis obsitis; mystacibus mediocribus, canis, ad basin nigricantibus; caudâ vix corpore breviore, suprà fusco-nigrá, subtùs sordidè albâ; pedibus anticis tarsisque flavescentibus

Description. Fur moderately long and loose; ears small; tail nearly equal to the body in length: general colour grey, with a wash of very pale yellow; chin, throat, and under parts of the body, white. Tail tolerably well clothed with hairs, those on the upper surface brown, and those on the under, whitish; on the sides are some yellowish hairs. Ears with yellow hairs on the inner side; tarsi pale yellow, toes white; muzzle and around the eye yellowish.

	Inches	*Lines*		*Inches*	*Lines*
Lĕngth from nose to root of tail	3	6[20]	Length of tarsus (claws included)	0	9½
of tail	2	1	of ear	0	4
from nose to ear	1	1			

Habitat, Santa Cruz and Port Desire. (*December*)

'Very common in long dry grass in the valleys of Port Desire.' D.

The skull is figured in Pl. XXXIII, fig. 5 *c*. Fig. 5 *a* represents the

[20] The dimensions given in the *Proceedings of the Zoological Society* were taken from a younger specimen than those here described, and there is an error in the length of the tail there given, which should be 1–10 instead of 2–10.

molars of the upper jaw; fig. 5 *b* those of the under jaw, and fig. 5 *d* represents the posterior molar of the under jaw when more worn.

It was with some hesitation that I described this as a distinct species in the Society's *Proceedings*. I have now re-examined the specimens, and still am / unable to satisfy myself whether they are varieties of *Mus xanthorhinus* or not. Both of *Mus canescens* and of *Mus xanthorhinus*, I have before me what I imagine to be an adult and a young specimen. The adult and the young of *M. xanthorhinus* agree in being of a *yellowish-brown* colour, and in having the muzzle and tarsi deep yellow; both specimens of *Mus canescens* are of a *grey* colour, with an indistinct yellow wash, the muzzle and tarsi being tinted with yellow, as in *M. xanthorhinus*. Besides this difference in tint, which, perhaps, is unimportant, *M. canescens* differs from *M. xanthorhinus* in having the head larger, the tail rather longer, and the fur less soft. The specimens of this animal are both from Patagonia; one of the specimens of *Mus xanthorhinus* was brought by Mr Darwin from Tierra del Fuego; and as the other formed part of Captain King's collection, it in all probability came from the same locality. As I only possess one skull, I cannot speak with certainty as regards the size of the head; the difference, however, in the stuffed specimens is considerable, and it is strange that each of the pairs should agree so perfectly, supposing the difference to be the work of the stuffer's hands.

18. MUS LONGIPILIS

PLATE XVI

Mus longipilis, *Waterh.*, *Proceedings of the Zoological Society of London* for February, 1837, p. 16.

M. suprà obscurè griseus, flavo lavatus; subtùs griseus; pedibus fuscis, unguibus longiusculis; auribus mediocribus; caudâ corpore breviore, suprà nigrescente, subtùs fuscescente; rhinario sub-producto: vellere longissimo, molli

Description. Fur very soft and silky, and extremely long – the ordinary fur of the back measuring nearly three quarters of an inch, and the longer hairs one inch in length; ears moderate; tail nearly as long as the body; muzzle much pointed; general colour grey, washed with yellow, the under parts pale grey, or greyish white; feet brown; ears and tail well clothed; the hairs on the inner side of the ears are chiefly of a yellow colour, those on the upper surface of the tail are brown black, those on the under part are dirty white; the hairs of the back are deep grey at the base, broadly annulated with yellow near the apex, and dusky at the apex; the longer hairs are greyish black; the hairs of the moustaches are dusky at the base, and whitish beyond that part; the claws are long, and but slightly curved; the / incisors are slender; those on the upper jaw are yellow, and those of the under yellow-white.

	Inches	*Lines*		*Inches*	*Lines*
Length from nose to root of tail	5	4	Length of tarsus (claws included)	1	0½
of tail	3	4	of ear	0	6½
from nose to ear	1	2			

Habitat, Coquimbo, Chile. (*May*)

This mouse is remarkable for the great length and softness of its fur, even among the species here described, most of which have very loose, long and soft fur.

The molars of the upper jaw are figured in Pl. XXXIII, fig. 6 *b* – molars of the lower jaw, fig. 6 *a*.

'Inhabits dry stony places, which character of country is general in this part of Chile.' D.

19. MUS NASUTUS

PLATE XVII – Fig. 2

Mus nasutus, *Waterh.*, *Proceedings of the Zoological Society of London* for February, 1837, p. 16.

M. suprà obscurè flavescenti-fuscus, ad latera fulvescens; subtùs obscurè fulvo tinctus: pedibus pilis obscurè fuscis tectis; unguibus longis; auribus mediocribus; caudâ corpore breviore, suprà fuscâ, subtùs sordidè albâ: rhinario producto

Description. Muzzle very long and pointed, ears small, tail shorter than the body, claws long and but slightly arched; inner, rudimentary toe of

PLATE XVI
Mus longipilis

the fore foot furnished with a pointed claw; fur moderate, and slightly glossy: general colour yellowish brown, of the sides of the body yellow, of the under parts pale yellow; the chin, throat and chest whitish: feet brown; ears well clothed with hairs, those on the inner side are most of them yellow, but some are black. All the fur is of a deep lead colour at the base; the hairs on the upper parts and sides of the head and body are broadly annulated with deep golden yellow near the apex, and blackish at the apex; on the upper parts long brownish black hairs are thickly interspersed with the ordinary fur, but on the side of the body they are less numerous, hence on this part / the yellow tint prevails; on the under parts of the body the hairs are broadly tipped with pale yellow, and in parts with white: the tail is but sparingly clothed with hairs, those on the upper surface are of a dark brown colour, and those on the under are pale brown. The incisors are very slender and of a very pale yellow colour.

	Inches	*Lines*		*Inches*	*Lines*
Length from nose to root of tail	5	2	Length of tarsus (claws included)	1	0½
of tail	2	8	of ear	0	5
from nose to ear	1	3			

Habitat, Maldonado, La Plata. (*June*)

The specific name *nasutus* has been applied to this mouse on account of its elongated and slender muzzle,[21] the tip of which extends nearly 4 lines beyond the upper pair of incisors: the rudimentary toe of the fore foot, instead of having the usual rounded nail, has a short pointed claw. Its fur is not so soft, nor yet so long as in many of the preceding species, and there is a greater admixture of yellow in its colouring. The claws appear to be adapted to burrowing.

The skull (which is not quite perfect) is figured in Pl. XXXIII, fig. 7 *a*, its length is 1 in. 3 lines. Fig. 7 *b*, represents the molars of the upper jaw, and fig. 7 *c*, those of the under jaw. The lower jaw, which is of a very slender and elongated form, is figured in Pl. XXXIV, fig. 10 *a*.

'Was caught in a small thicket on an open grassy plain, by a trap baited with a piece of bird. This mouse when alive possesses a marked character in the extreme acumination of its nose.' D.

[21] In *Mus longipilis* and *M. brachiotis* may be perceived an approach to this elongated form of the muzzle.

PLATE XVII

1. *Mus xanthorhinus* 2. *Mus nasutus*

20. MUS TUMIDUS

PLATE XVIII

Mus tumidus, *Waterh., Proceedings of the Zoological Society of London* for February, 1837, p. 15.

M. brunneus, nigro lavatus; rostro ad apicem, labiis, mento, gulâ, pectore, abdomineque albis; naso suprà nigrescente; auribus mediocribus rotundatis; corpore crasso; caudâ capite corporeque breviore, pilis nigricantibus, subtùs albescentibus prope basin, vestitâ; artubus pedibusque grisescentibus; vellere longo, molli; unguibus longis

Description. Body stout; head large; tail nearly as long as the head and body; / inner toe of the fore foot with a distinct, pointed claw; claws rather large, those of the fore feet but slightly arched. Fur rather long, and moderately soft; general tint of the upper parts of the body, brown, of the sides of the head and body, greyish, but with a yellow wash; the lower part of the sides of the body and of the cheeks, the tip of the muzzle, and the whole of the under parts, white; feet dirty white; ears densely clothed with short hairs, those on the inner side chiefly of an ashy-brown colour, and those on the outer side dusky; the hairs of the back are of a deep lead colour at the base, black at the tip, and annulated with yellow near the tip; the longer hairs, which are thickly interspersed, are totally black; on the under parts of the body the hairs are grey at the base, and broadly tipped with white; the upper surface of the muzzle is blackish; the moustaches are black; the incisors are yellow.

	Inches	*Lines*		*Inches*	*Lines*
Length from nose to the root of tail	6	9	Length of tarsus (claws included)	1	6
of tail	5	4	of ear	0	7
from nose to ears	1	8			

Habitat, Maldonado, La Plata. (*June*)

This species is about the size of *Mus Rattus*, but is stouter in its proportions; as in *Mus nasutus*, the thumb is furnished with a pointed claw. The molars of the lower jaw are figured in Pl. XXXIV, fig. 11 *a*.

'This rat was caught in so wet a place amongst the flags bordering a lake, that it must certainly be partly aquatic in its habits.' D.

PLATE XVIII

Mus tumidus

21. MUS BRAZILIENSIS

PLATE XIX

Rat du Brézil, *Geoff.*

M. suprà fuscus fulvo lavatus; lateribus capitis corporisque aequè ac abdomine auratis; gulâ pectoreque albis; pedibus pilis sordidè flavis tectis; auribus parvulis; caudâ caput corpusque ferè aequante; vellere longo, molli

Description. Head somewhat arched, and rather short; ears small; tail about equal in length to the head and body, measured in a straight line; tarsi large. Fur long, and rather soft; general colour deep golden yellow: on the upper surface of the head and the back, long glossy black hairs are thickly interspersed, and produce, with the admixture of the deep golden / colour of the ordinary fur, a dark brown tint; chin, throat, chest, and rump, white; the hairs covering the upper surface of the feet are of a dirty yellowish-white colour, and on the toes nearly white: ears densely clothed with longish hairs, those on the inner side chiefly of a deep golden colour, and those on the outer side brownish; the ears are partially hidden by the long fur of the head; tail sparingly clothed with hairs, above brown, and beneath brownish-white: the fur of the back is of a deep grey colour at the base, annulated with deep golden yellow near the apex, and blackish at the apex; the longer hairs are black; the hairs of the belly are pale grey at the base, and broadly tipped with golden yellow colour; the white hairs on the throat, chest, and rump are of an uniform colour – not tinted with grey at the root; the hairs of the moustaches are black: the incisors of the upper jaw are of a deep orange colour, and those of the lower jaw are yellow: the thumb nail is truncated.

	Inches	*Lines*
Length from nose to root of tail	8	6
of tail	7	9
from nose to ear	1	8
Length of tarsus	2	0
of ear	0	6½

Habitat, Bahia Blanca. (*September*)

This species is nearly equal in size to the common rat (*Mus*

PLATE XIX
Mus Braziliensis

decumanus). Of its skull[22] I possess but the anterior portion (see Pl. 33, fig. 3 *a* and 3 *b*): it appears to have been about the same size as that of *M. decumanus*, its proportions, however, are different: the nasal portion is broader and shorter, the ant-orbital outlet is rather smaller; the plate, forming the anterior root of the zygomatic arch, and which protects this outlet, has its anterior edge distinctly emarginated, and not nearly straight as in *M. decumanus* – the zygomatic arch is stouter, the space between the orbits is narrower, the palate is more contracted, the incisors are much broader, less deep from front to back, and have the anterior surface more convex; the molar teeth are larger; the lower jaw (see Plate 34, fig. 12 *a*) when compared with that of *Mus decumanus* also offers many points of dissimilarity; the principal differences consist in its greater strength, the comparatively large size and breadth of the articular surface of the condyles, the upright position of the coronoid process – a perpendicular line dropped from the apex of which would touch the posterior part of the last molar – and the great / extent of the *symphysis menti*. In the form of the incisors, the more contracted palate, the great extent of the *symphysis menti*, and in fact in most of the points of dissimilarity, between the skull of the present animal and that of *Mus decumanus*, here pointed out, it will be perceived, there is an approach made to the *Arvicolidae*.

The dimensions of the skull (so far as an imperfect specimen will allow of their being taken) are as follows:

	Inches	*Lines*
Distance between front of incisors (upper jaw), and the first molar tooth	0	8
Longitudinal extent of the three molars on either side, taken together	0	4½
Length of nasal bones	0	7¼
Length of incisive *foramina*	0	4¼
Width between orbits	0	2½
Length of *ramus* of lower jaw	1	1¼

Fig. 3 *c*, Pl. XXXIII, represents the molar teeth of the upper jaw. Fig 3 *d*, those of the upper jaw.

'This rat was caught at Bahia Blanca where the plains of Patagonia begin to blend into the more fertile region of the Pampas. It lived in holes amongst the tussocks of rushes, on the borders of a small, still,

[22] I am sorry to say the artist has not drawn this skull with his usual fidelity, a circumstance which I did not perceive until it was too late to make any alteration: it is too large, and the incisors are represented as projecting forwards too much; they are in the original so nearly at right angles with the upper surface of the skull that but a very small portion of them is seen, which it is viewed, as represented at fig. 3 *a*.

brook; in its manner of diving and aquatic habits it closely resembled the English water-rat (*Arvicola amphibia*).' D.

When at Paris I examined what I believe to be the original *Mus Braziliensis*, since the specimen was labelled '*Rat de Brazil St Hilaire, 1818*'. It agrees perfectly with the present animal excepting in being rather smaller, the length from the nose to the tail being 7 inches and 4 lines – the length of the tail is 7 inches 9 lines, and that of the tarsus is 1 inch 11 lines; this difference in the length of the body may arise from difference of age, or even of sex. In the Paris Museum I saw what appeared to me to be a variety of the same species in which the under parts of the body are white.

I have been minute in my description of the *Mus Braziliensis*, since it is confounded by Desmarest, Fischer and Lesson with the *Rat troisieme* or *Rat Angouya* of Azara, which I believe to be a very different animal. The description given by the authors just mentioned are taken from Azara, who gives the following characters to distinguish the Rat Angouya: 'Du museau à la queue, et sur les côtés du corps tout est brun-cannelle, parceque les poils ont une petite pointe cannelle; puis, ils sont obscurs et enfin blanc vers las peau. Toute la partie inférieure de l'animal est blanchâtre, plus claire sous la tête, et plus foncée entre les jambes de devant; le pelage est doux, très-serré, et le poil, qui est à la racine de l'orielle, cauche le conduit de celle-ci.' /

	Inches	*Lines*		*Inches*	*Lines*
Length from nose to root of tail (English measure)	6	0	Length of ears	0	9¾
of tail	6	6½	of tarsus (the claws included)	1	3¼

It appears from this description that the *Mus Angouya* is a smaller animal, and differs both in colouring and proportions from the *Mus Braziliensis*. Brandt has figured and described a rat under the name of *Mus Angouya*, which in many respects agrees better with Azara's description; there are, however, discrepancies in the dimensions.

22. MUS MICROPUS

PLATE XX

Mus micropus, *Waterh.*, *Proceedings of the Zoological Society of London* for February, 1837, p. 17.

M. suprà fuscus; subtùs cinerescenti-albus, pallidè flavo tinctus; pedibus pilis sordidè albis tectis, antipedibus parvulis; auribus parvulis; caudâ, quoad longitudinem, corpus ferè aequante, suprà fuscâ, subtùs sordidè albâ

Description. Form stout, ears rather small, tail nearly equal to the body in length, fur very long and moderately soft, general colour of the upper parts of head and body, brown; of the sides of the body greyish, faintly washed with yellow, of the under parts greyish white, faintly tinted with yellow; hair covering the upper surface of the feet dirty white; on the tarsus there is a very slight yellow tint; ears well clothed with hairs, those on the inner side chiefly of a yellow colour; tail above, dusky brown; beneath dirty white: hairs of moustaches black at the base and greyish at the apex; incisors pale yellow: hairs of the back deep grey at the base, annulated with brownish yellow near the apex, and dusky at the apex; longer hairs dusky black; hairs of the belly deep grey at the base and broadly tipped with yellowish white.

	Inches	*Lines*		*Inches*	*Lines*
Length from nose to root of tail	6	0	Length of tarsus (claws included)	1	0¾
of tail	3	8	of ear	0	6
from nose to ear	1	4			

Habitat, Santa Cruz, Patagonia. (*April*) /

The molars of the upper jaw are figured in Pl. XXXIV, fig. 13 *a*, and those of the lower jaw, fig. 13 *b*.

'Caught in the interior plains of Patagonia in lat. 50°, near the banks of the Santa Cruz.' D.

PLATE XX
Mus micropus

23. MUS GRISEO-FLAVUS

PLATE XXI

Mus griseo-flavus, *Waterh.*, *Proceedings of the Zoological Society of London* for February, 1837, p. 28.

M. suprà griseus flavo-lavatus, ad latera flavus, subtùs albus; pedibus albis; auribus magnis et ferè nudis; caudâ caput corpusque ferè aequante, suprà fusco-nigricante, subtùs albâ; vellere longo, molli

Description. Ears large; tail rather shorter than the head and body taken together; tarsi slender, and moderately long; fur long and very soft; general tint of the upper parts of head and body greyish, washed with brownish yellow; on the sides of the body a palish yellow tint prevails; feet, chin, throat, and under parts of body pure white; tail rather sparingly clothed with hairs, those on the apical portion rather long, and forming a slight pencil at the tip; on the upper side and at the tip of the tail the hairs are brown, on the under side they are dirty white; the ears are very sparingly clothed with minute brownish yellow hairs internally; externally, on the fore part, the hairs are rather longer and of a brown colour; the upper incisors are orange, and the lower incisors are yellow; the hairs of the moustaches are long, and of a black colour; the hairs of the back are deep grey at the base, brownish at the tip, and annulated with pale brownish yellow near the tip; the longer hairs are brown; the hairs of the belly are white externally, and grey at the base; on the throat the hairs are white to the root.

	Inches	*Lines*		*Inches*	*Lines*
Length from nose to root of tail	6	8	Length of tarsus (claws included)	1	2½
of tail	5	6	of ear	0	8
from nose to ear	1	4½			

Habitat, Northern Patagonia. (*August*)

The molars of the upper jaw are figured in Pl. XXXIV, fig. 15 *a*, and those of the lower jaw, fig. 15 *b*.

'Inhabits the dry gravelly plain, bordering the Rio Negro.' D. /

PLATE XXI

Mus griseo-flavus

24. MUS XANTHOPYGUS

PLATE XXII

Mus xanthopygus, *Waterh., Proceedings of the Zoological Society of London* for February, 1837, p. 28.

M. suprà pallidè brunneus flavo-lavatus, ad latera flavescens, subtùs albus; capite griscescente; natibus flavis; pedibus albis; auribus majusculis pilis, albis et flavis intermixtis obsitis; caudâ quoad longitudinem, corpus ferè aequante, suprà nigricante, subtùs albâ; vellere longo et molli; mystacibus perlongis albescentibus, ad basin nigris

Description. Ears rather large, tail rather longer than the body, tarsi moderately long and somewhat slender: fur long and very soft: prevailing tint pale yellow; on the back there is a brownish hue owing to the long hairs, which are thickly interspersed with ordinary fur, being of that colour: in the region of the tail the hairs are of a rich yellow colour; the tip of the muzzle is white, the feet, chin, throat and the whole under parts of the body are white; on the chest and belly a faint yellowish hue is observable: the tail is well clothed with tolerably long hairs, those on the apical portion are the longer, on the upper side of the tail they are of a brown colour, and on the under side they are pure white: the ears are well clothed with tolerably long hairs, those on the inner side are of a pale yellowish colour, externally on the fore part they are brown, and on the hinder part they are yellowish white: the hairs of the moustaches are numerous and very long; some of them are white, but the greater portion are brownish black at the base and whitish at the apex: the upper incisors are yellow, and the lower are yellow-white: the hairs of the ordinary fur on the back are grey at the base, brownish at the tip, and very pale yellow near the tip: the hairs on the belly are grey at the base and white externally.

	Inches	*Lines*		*Inches*	*Lines*
Length from nose to root of tail	5	3	Length of tarsus (claws included)	1	1
of tail	3	10	of ear	0	7
from nose to ear	1	3			

There are three specimens of the present species in Mr Darwin's collection; two of them were caught when shedding their fur, and

PLATE XXII

Mus xanthopygus

having lost the longer black hairs, have the upper parts of the body of a paler colour; their general tint is very pale, and may be described as grey, with a wash of pale yellow. /

This species is closely allied to the last, but differs in being rather smaller, in having smaller ears which are well clothed with hair, and not sparingly furnished as in *Mus griseo-flavus*, and in having a shorter tail which, like the ears, is more densely clothed with hairs; in the structure of the molar teeth there are also differences which will be better understood by comparing the drawings. Fig. 16 *a*, Pl. XXXIV, represents the molars of the upper jaw, and 16 *b*, those of the lower jaw.

'Extremely abundant in the coarse grass and thickets in the ravines at Port Desire and Santa Cruz: was caught in a trap baited with cheese.' D.

25. MUS DARWINII

PLATE XXIII

Mus Darwinii, *Waterh., Proceedings of the Zoological Society of London* for February, 1837, p. 28.

M. suprà pilis pallidè cinnamomeis et nigrescentibus intermixtis; ante oculos cinerascentibus; genis, lateribus corporis, et caudâ prope basin, pallidè cinnamomeis; partibus inferioribus pedibusque albis; auribus permagnis; caudâ caput corpusque ferè aequante, suprà fusco-nigricante, subtùs albâ

Description. Form robust; ears immensely large; tail nearly equal in length to the head and body taken together; fore feet very small; tarsi moderate; fur very long and soft; general tint of the upper parts pale cinnamon yellow; on the rump a richer yellow hue prevails, and on the back there is a brownish tint, owing to the interspersed long hairs being of that colour; the upper surface of the head is greyish; the cheeks, like the sides of the body, are of a delicate yellow colour, faintly clouded with brown; the sides of the muzzle, lower part of the cheeks and sides of the body, and the whole under parts, are pure white; the feet and tail are also white, if we except the upper surface of the latter, which is dark brown; the yellow tint of the sides of the body is extended downwards on the outer side of the fore legs and on the back

PLATE XXIII

Mus Darwinii

of the hinder legs; the ears are but sparingly furnished with hair, excepting on the fore part, externally, where they are of a brownish colour; the minute hairs which cover the remaining parts of the ear are very pale; the tail is well clothed with hairs; the hairs of the moustaches are numerous and very long; they are for the most part blackish at the base, and grey at the apex; the incisors are rather slender, the upper pair are an orange colour, and the lower, yellow; the hairs of the ordinary fur of the back are grey at / the base, broadly annulated with pale cinnamon yellow near the apex, and brownish at the apex; the hairs of the belly are deep-grey at the base, and white externally, those on the throat are pale grey at the base.

	Inches	*Lines*		*Inches*	*Lines*
Length from nose to root of tail	6	0	Length of tarsus (claws included)	1	1½
of tail	4	9	of ear	0	11¾
from nose to ear	1	4½	Width of ear	1	0½[23]

Habitat, Coquimbo, Chile. (*May*)

This species is evidently allied to the two preceding; and perhaps the 'Rat quatrieme, ou Rat oreillard' of Azara (*Mus auritus*, Desm.), will form one of this little group. The molar teeth of the upper jaw are figured in Pl. XXXIV, fig. 17 *a* – those of the lower jaw, fig. 17 *b*.

'Inhabits dry stony places.' D.

26. MUS GALAPAGOENSIS

PLATE XXIV

M. suprà fuscus, flavo-lavatus, ad latera flavescens, subtùs albus: pedibus pilis sordidè albis tectis: aùribus mediocribus; caudâ, quoad longitudinem, caput corpusque ferè aequante: vellere longo

Description. Ears moderate, slightly pointed; tarsi moderate; tail slender, nearly as long as the head and body; fur long, and not very

[23] It is not easy to measure the *width* of the ears in these animals: upon measuring with a thread over the curve of the outer side I have found the width of the ears of the present animal to be as above given – the dimension slightly exceeding that stated in the *Proceedings of the Zool. Soc.*

PLATE XXIV
Mus Galapagoensis

soft; upper parts of the body of a brownish hue, a tint produced by the admixture of black and palish yellow hairs; on the sides of the body the longer black hairs are less abundant, and the prevailing colour is yellow; under parts of the body white, with a very faint yellow tint; feet furnished above with dirty white hairs; ears rather sparingly clothed with hairs, those on the inner side of a yellow colour, and those on the outer side dusky; tail above brown, and beneath whitish; the hairs of the moustaches black; the incisors deep yellow; the hairs on the back are deep grey at the base, broadly annulated with palish yellow near the apex, and blackish at the apex; the longer hairs black; on the belly the hairs are grey at the base, and broadly tipped with yellowish white. /

	Inches	*Lines*		*Inches*	*Lines*
Length from nose to root of tail	6	0	Length of tarsus (claws included)	1	2
of tail	4	9	of ear	0	7
from nose to ear	1	3¾			

Habitat, Chatham Island, Galapagos Archipelago, Pacific Ocean. (*October*)

This species is less than *Mus Rattus*. The upper parts of the body have a slightly variegated appearance.

The skull of *Mus Galapagoensis* (Pl. XXXIII, fig. 8 *a*) is rather smaller than that of *M. Rattus*, the nasal portion is proportionately longer, the cranial shorter, and the interparietal bone is smaller, especially in antero-posterior extent; its length is 15 lines, and its breadth is 8⅛ lines. The lower jaw is figured in Pl. XXXIV, fig. 14 *a*. Fig. 8 *b* of Pl. XXXIII, represents the molars of the upper jaw, and fig. 8 *c*, those of the lower jaw.

'This mouse or rat is abundant in Chatham Island, one of the Galapagos Archipelago. I could not find it on any other island of the group. It frequents the bushes, which sparingly cover the rugged streams of basaltic lava, near the coast, where there is no fresh water, and where the land is extremely sterile.' D.

27. MUS FUSCIPES

PLATE XXV

M. suprà fusco-nigrescens, subtùs griseus; pedibus fuscis; auribus mediocribus, caudâ, quoad longitudinem, caput corpusque ferè aequante: vellere longissimo, molli

Description. Form stout; ears moderate; tail equal to the body in length; tarsi moderate; fur very long. General tint of the upper part and sides of the head and body blackish brown with an admixture of grey; of the under parts greyish white; feet brown, the hairs greyish at the tip: tail black and but sparingly clothed with short bristly hairs: ears rather sparingly clothed with hairs, which are for the most part of a brownish grey colour. The ordinary fur of the back is about ¾ of an inch in length and very soft – of a deep grey colour, broadly annulated with brownish yellow near the tip and blackish at the tip: the longer hairs which are black, measure upwards of 1¼ inches in length. The upper incisors are of an orange colour and the lower are black.

	Inches	*Lines*		*Inches*	*Lines*
Length from nose to root of tail	6	6	Length of ear	0	6½
of tail	4	3	of tarsus (claws included)	1	1
from nose to ear	1	6			

Habitat, Australia, King George's Sound. (*March*) /

Mammalia not belonging to the order *Marsupiata* are rare in the Continent of Australia. Besides the Dog, we are acquainted with none excepting a few species of Rodents, and these all belong to the family *Muridae*.

The present animal adds one to the limited number already known: in the Museum of the Zoological Society there is another species, the characters of which I will point out in the next description.

Mus fuscipes is remarkable for the great length and softness of its fur, and the brown colour of its feet: it is rather less than *Mus Rattus*, and of a stouter form. Not having had an opportunity of examining the molar teeth and the cranium of this animal, I cannot be positive that it is a species of the genus *Mus*; in external characters and the form of the incisor teeth, however, it agrees perfectly with the animals of that genus.

'This animal was caught in a trap baited with cheese, amongst the bushes at King George's Sound.' D.

28. MUS GOULDII

M. vellere longo, molli, ochraceo, pilis nigricantibus adsperso, his ad latera rarioribus: corpore subtùs, pedibusque albis: auribus majusculis: caudâ, capite corporeque paulo breviore

Description. Ears rather large and slightly pointed, tarsi slender and tolerably long; tail about equal in length to the body and half the head; fur long and soft; general colour pale ochreous yellow; on the back there are numerous long black hairs interspersed with the ordinary fur, which gives a darker hue and somewhat variegated appearance to that part; feet, chin, throat, and the whole under-parts of the body white; ears brown, sparingly clothed with minute yellow hairs, both externally (excepting on the forepart, where they are brownish) and internally; tail brownish above, and yellowish white beneath; the hairs of the moustaches long, and of a brown colour; upper incisors deep orange, lower incisors yellow; claws white. The hair of the back is of a deep lead colour at the base, pale ochre near the apex, and dusky at the apex; the longer hairs are black; the hairs of the belly are deep grey at the base and broadly tipped with white.

	Inches	*Lines*		*Inches*	*Lines*
Length from nose to root of tail	4	8	Length of tarsus (claws included)	1	0½
of tail	3	6	of ear	0	7
from nose to ear	1	0½			

Var. β. General colour of the fur pale ochreous yellow, the feet, under side of the tail and the whole of the under parts, as well as the lower portion of the / sides of the body, white; hairs of the back palish grey at the base, those of the belly indistinctly tinted with very pale grey at the roots; ears and moustaches pale brown.

Habitat, New South Wales.

PLATE XXV

Mus fuscipes

This species is about half-way between *Mus Rattus* and *Mus musculus* in size, and is remarkable for its delicate colouring. The molar teeth are figured in Pl. XXXIV; fig. 18 *a*, represents the molars of the upper jaw, and fig. 18 *b*, those of the lower.

GENUS: REITHRODON[24]

Dentes primores 2/2; inferioribus acutis, gracilibus, et anticè laevibus; superioribus gracilibus, anticè longitudinalitèr sulcatis

Molares untrinque 3/3 radicati; primo maximo, ultimo minimo: primo superiore plicas vitreas duas externè et internè alternatim exhibente; secundo, et tertio, plicas duas externè, internè unam: primo inferiore plicas vitreas tres externè, duas internè; secundo, plicas duas externè, unam internè; tertio unam externè et internè, exhibentibus

Artus inaequales: antipedes 4-dactyli, cum pollice exiguo: pedes postici 5-dactyli, digitis externis et internis brevissimis

Ungues parvuli et debiles. Tarsi subtùs pilosi

Cauda mediocris, pilis brevibus adpressis instructa

Caput magnum, fronte convexo: oculis magnis: auribus mediocribus

The present genus according to my views belongs to the family *Muridae*. The modifications of structure which have led me to separate it from the genus *Mus* are as follows:

External characters. The most conspicuous points of distinction between the external characters of *Reithrodon* and *Mus* (if we regard *M. rattus*, *M. decumanus* or *M. musculus* as typical examples of that genus), consist in the arched form of the head, the large size of the eyes, the stout form of the body, and the upper incisors being grooved. The ears, tail and feet are more densely / clothed with hairs, and the tarsus is covered with hair beneath – at least the hinder portion.

Cranium. The skulls of the species of the present genus differ from those of the species of *Mus* in being proportionately shorter and broader, and more arched; the facial portion of the skull is larger, compared with the cranial, the space between the orbits is narrower, and the orbits are larger; the palate is narrower and the incisive

[24] Ρειθρος, a channel; Οδον, a tooth.

foramina are more elongated and larger. The pterygoids approximate anteriorly, so that the posterior *nares* are greatly contracted. As in the genus *Mus* the anterior root of the zygomatic arch is directed upwards from the plane of the palate, and forwards in the form of a thin plate, protecting an opening behind, which leads into the nasal cavity, and also forming the outer boundary both of the ant-orbital foramen, and a second opening whose outlet is directed upwards. This thin plate, however, is narrower than is usually found in the genus *Mus*. The most striking differences observable in the lower jaw consist in the smaller size of the coronoid process, and its being curved outwards; the condyloid process is narrower, and the angle of the jaw, or descending ramus, approaches more nearly to a quadrate form – the posterior edge of the jaw is more deeply emarginated.

Dentition. The incisors are narrow and compressed as in the genus *Mus*, but they are less deep from front to back; those of the upper jaw (Plate 33, fig. 2 *b*) have each a distinct longitudinal groove, which is situated nearer to the outer than to the inner edge of the tooth. Close to the inner edge of each of these teeth an indistinct second longitudinal groove may be seen by means of a lens. The lower incisors are nearly equal in width to the upper.

The crowns of the molar teeth in the young *Reithrodon* are higher than in *Mus*, and they are rootless; in the adult animal, however, they possess distinct roots. The folds of enamel form sigmoid flexures, are closely approximated to each other, and those of the opposite sides of the tooth meet.

1. REITHRODON CUNICOLOÏDES

PLATE XXVI

Reithrodon cuniculoïdes, *Waterh., Proceedings of the Zoological Society of London* for February, 1837, p. 30.

R. suprà griseus, flavo-lavatus, pilis nigris intermixtis; abdomine gulâque pallidè flavis; natibus albis; pedibus albis; auribus mediocribus, intùs pilis flavis, extùs pilis pallidè flavis, obsitis; pone aures, notâ magnâ albescenti-flavâ; caudâ corpore breviore, suprà pallidè fuscâ, subtùs albâ

Description. Head rather large and arched; ears moderate; tail nearly as long / as the body; tarsi rather long; fur long and very soft. General tint of the upper parts of the body greyish brown, with a considerable admixture of yellow; of the sides of the body greyish tinted with yellow; the lower portion of the cheeks, and the lower half of the sides of the body are of a delicate yellow colour; the under parts of the head and body are yellowish white; the fore part of the thighs is whitish; the rump, feet, and tail are white, excepting the upper surface of the latter, which is brown; behind each ear there is a patch of yellowish white hairs. The ears are tolerably well-clothed with hairs; those on the inner side are for the most part of a yellow colour, but towards the posterior margin they are brown; externally, the hairs are also yellow, excepting on the fore part, where they are dusky brown. The hairs of the moustaches are very long and numerous; black at the base, and greyish at the apex. The feet are well clothed with hairs which cover and nearly hide the claws; the under side of the tarsus is clothed with greyish brown hairs. The tail is well clothed with tolerably long hairs which completely hide the scales. The hairs on the back are of a deep grey colour at the base, broadly annulated with yellow near the apex, and dusky at the apex: the longer hairs are black: on the throat and belly the hairs are deep grey at the base, and broadly tipped with pale yellow – towards the cheeks and sides of the body with a deeper yellow. The incisors are yellow.

	Inches	*Lines*		*Inches*	*Lines*
Length from nose to root of tail	6	5	Length of tarsus (claws included)	1	4 1/1
of tail	3	3½	of ear	0	7
from nose to ear	1	4			

Habitat, Patagonia. (*April and January*)

In the arched form of the head this little animal bears considerable resemblance to a young rabbit, a resemblance which has struck almost all who have seen it, I have therefore applied to it the specific name *Cuniculoïdes*. The skull is figured in Pl. XXXIII, fig. 2 *a*, its dimensions are as follows:

	Inches	*Lines*
Total length	1	4
Width	0	10
Length of nasal bones	0	7
Length of incisive foramina	0	4 3/1
Distance between the outer surface of the incisors and the front molar upper jaw	0	5
Longitudinal extent of the three molars of the upper jaw	0	3 3/1
Length of a ramus of the lower jaw, without the incisor	0	9 3/1

PLATE XXVI

Reithrodon cuniculoïdes

The molar teeth of the upper jaw are figured in Pl. XXXIII, fig. 2 *c* and / 2 *e*; of the lower jaw, fig. 2 *d*. Fig. 2 *b*, represents the incisors of the upper jaw magnified. Fig. 21 *a*, Pl. XXXIV, represents the skull, viewed from beneath, fig. 21 *b*, is the side view of the same, and fig. 21 *c*, is the lower jaw.

'Specimens were procured at Port Desire, St Julian, and Santa Cruz; at this latter place they were caught in numbers (in traps baited with cheese), both near the coast and on the interior plains. A specimen from Santa Cruz weighed 1336 grains. In the early part of January, there were young individuals at Port St Julian.' D.

2. REITHRODON TYPICUS

Reithrodon typicus, *Waterh.*, *Proceedings of the Zoological Society of London* for February 1837, p. 30.

R. vellere suprà pilis flavescenti-fuscis et nigrescentibus intermixtis composito; regione circa oculos, genis, lateribusque corporis auratis, pilis pallidè fuscis intermixtis; partibus inferioribus auratis; rhinario ad latera flavescenti-albo; auribus magnis, intùs pilis flavis, extùs flavis et fuscis, indutis; caudâ suprâ pallidè fuscâ, subtùs sordidè albâ; pedibus albis

Description. Ears large; tarsi moderate; fur moderately long; general tint of the upper parts brown – of the upper surface of the head blackish; on the cheeks and flanks a rich yellow tint prevails; the under parts of the head and body are bright yellow; the feet are white; the tail is brownish above and dirty white beneath. The ears are tolerably well clothed with hairs, and these are of a yellowish colour, excepting on the fore part, externally, where they are brown; the tarsi are covered beneath with greyish brown hairs; the hairs of the moustaches are numerous and moderately long, black at the base and greyish at the apex. The hairs of the back are deep grey at the base, broadly annulated with yellow near the apex, and black at the apex; on the upper surface of the head the hairs are very narrowly annulated with yellow, hence a blackish hue prevails. The longer hairs on the back are black; the hairs of the throat and belly are grey at the base, and broadly tipped with yellow. The incisors are yellow.

	Inches	Lines		Inches	Lines
Length from nose to root of tail	6	0	Length of tarsus (claws included)	1	2½
of tail		?[25]	of ear	0	8½
from nose to ear	1	4½			

Habitat, Maldonado, La Plata. (*June*) /

This species is of a darker colour than the last, its ears are much larger and the tarsi are shorter. It has the same rabbit-like appearance. The molar teeth of the lower jaw are figured in Pl. XXXIII, fig. 4 *a*.

'This mouse, when alive, from its very large eyes and ears, had a singular appearance, somewhat resembling that of a little rabbit. It frequents small thickets in the open grassy savannahs near Maldonado, and was caught with facility by means of traps baited with cheese.' D.

3. REITHRODON CHINCHILLOIDES

PLATE XXVII

R. vellere longissimo et mollissimo; corpore suprà et ad latera cinereo, flavescenti-fusco lavato, subtùs flavescenti-albo; caudâ corpore breviore, suprà fuscâ, subtùs albâ: auribus parvulis: tarsis mediocribus

Description. Ears small; tail shorter than the body; tarsus moderate; fur long and extremely soft. General hue of the upper parts of the head and body ashy-brown; the lower part of the cheeks and sides of the body are of a delicate yellow colour; the under parts of the head and body and the rump are cream colour. The ears are blackish;[26] the tail is tolerably well clothed with longish hairs, which are, however, not so thickly set as to hide the scales – on the upper side they are blackish brown; on the sides and beneath they are white. The feet are white. All the fur is of a deep grey colour at the base; the hairs of the back are of a very pale yellow colour (almost white) near the tip, and brown at the tip; the longer hairs are black at the apex. The incisors are yellow; the hairs of the moustaches are numerous and very long – some of them are whitish, and others are black at the root, and grey at the apex.

[25] The tail is imperfect.
[26] They are naked, but I suspect the hair has been rubbed off.

	Inches	Lines		Inches	Lines
Length from nose to root of tail	5	0	Length of tarsus (claws included)	1	0
of tail	2	4	of ear	0	5½
from nose to ear	1	2			

Habitat, South shore of the Strait of Magellan, near the Eastern entrance.

This little animal was preserved in spirit, and has since been mounted, it is / probable, therefore, that the colours have been slightly changed. It is of a smaller size than either of the preceding species. Its fur is long, extremely soft, and somewhat resembles that of the Chinchilla. The ears are smaller, and the tail is shorter, and less densely clothed with hairs than in *Reithrodon cuniculoides*. The skull (see Pl. XXXIV, fig. 20 *a*, 20 *b*, and 20 *c*) differs in many respects from that of the species last mentioned. It is of a smaller size, the nasal portion is proportionately shorter and narrower, the incisive foramina are shorter; the pterygoid processes do not approximate so nearly at their base, and the pterygoid fossae are very shallow, whereas in *R. cuniculoïdes* they are deep. In the skull of the animal just mentioned there are two distinct longitudinal grooves on the palate, which extend backwards from the incisive foramina, and terminate in two rather large and deep excavations: these excavations are in the palatine bone, and situated between the last molar teeth; they are separated from each other by a narrow longitudinal, elevated ridge; a narrow ridge also separates them from the pterygoid fossae. At the bottom of each of these hollows are several minute foramina, and in front of them there are two larger longitudinal foramina. In *R. chinchilloides*, the longitudinal grooves on the palate and the posterior hollows are shallow, and consequently much less distinct; the pterygoid fossae are very nearly on the same plane as the palate, and are indicated only by a very slight depression. The incisor teeth are broader than in *R. chinchilloides*, and the molar teeth are proportionately smaller. The thin plate which forms the anterior root of the zygomatic arch is deeply emarginated in front of *R. cuniculoïdes* (see Pl. XXXIV, fig. 21 *b*); but in *R. chinchilloides*, the anterior margin of this plate is nearly straight (see Pl. XXXIV, fig. 20 *c*).

In the form of the lower jaw of the two animals under consideration there are differences which will be more clearly understood upon comparing the figures. I will therefore merely notice one remarkable

PLATE XXVII

Reithrodon chinchilloides

character which is found in *R. cuniculoïdes*, and that is, that the condyloid process is rather deeply concave on the inner side, a character which does not exist in *R. chinchilloides*, nor do I recollect having observed it in any other Rodent.

The principal dimensions of the skull of *R. chinchilloides*, are as follows:

	Inches	*Lines*
Total length	1	2
Width	0	8½
Length of nasal bones	0	6⅓
Length of incisive foramina	0	4
Distance between the outer surface of the incisors and the first molar tooth, upper jaw	0	4½
Longitudinal extent of the three molars of the upper jaw, taken together	0	2¾
Length of a ramus of the lower jaw without the incisor	0	8 /

General Observations upon the foregoing Species of Muridae

In the foregoing descriptions I have endeavoured to convey an idea of the characters of the species of mice submitted to me for examination and description, by Mr Darwin: there are, however, some points upon which I have been silent in my descriptions. I allude to the characters observable in the dentition. I have omitted to notice the various modifications in the structure of the molar teeth, because I found it would lengthen the descriptions to no good purpose, inasmuch as of almost all the species I have made outlines of the molars, which will convey a more clear idea than any verbal description can do.

Upon an inspection of the Plates, it will be seen, that by far the greater portion of the teeth figured, may be referred to one particular type of form or pattern, and that this pattern does not agree with that observed in the molars of *Mus Rattus*, *M. decumanus*, or *M. musculus*, whilst these three species agree essentially with each other.

In the young Black Rat (*Mus Rattus*), before the teeth are worn, the two anterior molar teeth, on either side of the upper jaw, present three longitudinal rows of tubercles, a central series of larger tubercles, and on each side of these, a row of smaller ones. The front molar has three of the larger tubercles arranged along the middle of the tooth; three smaller ones on the outer side, and two, on the inner side. The second molars have two central tubercles, two outer, and two inner ones. The posterior molar is nearly round, the body of the tooth consists of three principal tubercles, and one small tubercle, situated on the inner and anterior portion of the tooth.

The corresponding teeth in the young of *Mus bimaculatus* present a very different appearance; the molars, instead of having three longitudinal rows of tubercles, have only two. An idea of the appearance of these teeth may be formed by removing the inner row of tubercles from the molars of *Mus rattus*. We should then have, as in *Mus bimaculatus*, molars of a narrower form, the first tooth presenting six tubercles, the second, four; and the posterior tooth devoid of the small inner lobe; the opposing tubercles of each tooth, however, in *M. bimiculatus*, are of equal size.

The molars of the lower jaw of *Mus bimaculatus* agree with those of *M. Rattus* as to the number of tubercles which they possess; they are, however, proportionately longer and narrower, and, when a little worn, these teeth, as well as those of the upper jaw, differ considerably from those of *M. Rattus*. In the last named animal, when the molars are slightly worn, the ridges of enamel run completely across the tooth, as in Figs 18 and 19, Pl. XXXIV. Such is not the case / in *M. bimaculatus* at any age. As soon as the molar teeth are worn, the folds of enamel penetrate the body of the tooth on each side, and those of one side alternate with those of the other – in fact, they very nearly resemble those of the *Hamsters* (*Cricetus*).

I have selected the molar teeth of *Mus Rattus* and *M. bimaculatus* for comparison, since I happened to possess specimens displaying both the young and adult states of each. But had I selected, on the one hand, almost any of the species brought from South America by Mr Darwin, and, on the other hand, the *Mus musculus* or *M. decumanus*, I should have had to point out the same distinctions – the former agreeing in dentition with *M. bimaculatus*, and the latter with *M. Rattus*.

The differences pointed out, between the molar teeth of *Mus Rattus* and those of *M. bimaculatus*, I cannot but consider as important, since all the Old World species of *Mus* which I have yet had an opportunity of examining (and they are numerous) agree essentially with the former, whilst the only *Mus* from S. America (excepting *M. Musculus* and *M. decumanus*, which are carried in ships to all parts of the world) in which I have as yet found molar teeth like those of *M. Rattus*, is the *Mus Maurus*, and this it has been stated is possibly a variety of *M. decumanus*.

Although as yet I have not met with species in the Old World possessing the characters of the South American *Muridae*, among those of North America, several have come under my observation. The *Mus leucopus*, *Sigmodon hispidum*, and the species of *Neotoma* certainly belong to the same group,[27] as does also the species of the Galapagos Islands, described in this work under the name *Galapagoensis*.

These considerations have induced me to separate the South American mice from those of the Old World – or rather from that group of which *M. decumanus* may be regarded as the type – and to place them, together with such North American species as agree with them in dentition, in a new genus bearing the name *Hesperomys*.[28]

Whether this group be confined to the Western hemisphere or not, I will not venture to say, but I think I may safely affirm that that portion of the globe is their chief metropolis.

The species of the genus *Hesperomys*, which depart most from the type – whose dentition is least like figs 5 *a* and 5 *b*, Pl. XXXIII, or 6 *a* and 6 *b* of the / same Plate – recede still farther from the genus *Mus*, and approach more nearly (as regards the dentition) to the *Arvicolidae*. Among the species here described I may mention as examples, *M. griseo-flavus*, *M. xanthopygus*, and *M. Darwinii*; see the molar teeth figured in Pl. XXXIV, figs 15, 16, and 17 – and among the North American species, those constituting the genus *Neotoma*. The latter make by far the nearest approach to the *Arvicolidae* of any which have yet come under my observation, not only in the dentition, but in the form of the skull and the large size of the coronoid process of the lower jaw; there is, nevertheless, a tolerably well marked line of distinction between the crania of the *Arvicolidae* and *Neotoma*.

The skulls of the animals belonging to the genera *Castor*, *Ondatra*, *Arvicola*, *Spalax*, and *Geomys*, which constitute the principal groups of the family *Arvicolidae*, when compared with those of the family *Muridae*, present, among others, the following distinctive characters.

The temporal *fossae* are always much contracted posteriorly, by the great anterior and lateral development of the temporal bones; the plane of the inter-molar portion of the palate is below the level of the anterior portion; the coronoid process of the lower jaw is very large, the articular portion of the condyloid process is proportionately broad; the descending ramus, or posterior coronoid process, is so situated that its upper portion terminates considerably above the level

[27] I am acquainted with seven North American Species of *Muridae*, all of which possess the dentition of *Hesperomys*.

[28] Εσπερος, West, and Μυς.

of the crowns of the molars; this same process is generally[29] directed outwards from the plane of the horizontal ramus. The incisor teeth of the *Arvicolidae* differ from those of the *Muridae* in being proportionately broader and less deep from front to back – they are not laterally compressed as in *Mus*. The molar teeth are rootless,[30] and the folds of enamel are the same throughout the whole length of the tooth; whereas in *Mus* they enter less and less deeply into the body of the tooth as we recede from the crown, and towards the base of the visible portion (the tooth being in its socket) the indentations of the enamel are obliterated.

Now in the species of *Hesperomys*, the molar teeth are always rooted, and in the form of the skull and the lower jaw they agree with the *Muridae*, and do not / present the characters above pointed out as distinguishing the *Arvicolidae*, and as regards the cranium and lower jaw, it is only in the genus *Neotoma* that any approach is evinced.

Of the various groups of the order *Rodentia* found in South America, the *Sciuridae*, so far as I am aware, are chiefly confined to the more northern parts, and do not occur in the most southern; the *Myoxidae*, *Gerboidae*, and *Arvicolidae* are wanting. The species of the family *Muridae* belong to different sections to those of the Old World. Of the *Leporidae* I am acquainted only with one well established species – the *Lepus Braziliensis*, which however is not found 'in tota America Australi', as Fischer says, there being no Hare yet found in the more southern parts, where the *Cavies* and *Chinchillas* appear to take their place. The remaining South American Rodents – certain species of *Hystricidae*, the genera, *Echimys*, *Dasyprocta*, *Caelogenys* and *Myopotamus*, together with the *Octodontidae* and *Chinchillidae*, all possess a peculiar form of skull and of the lower jaw (more or less approaching to figs 1, Pl. XXXIII, and figs 23, Pl. XXXIV) which I have described in the *Magazine of Natural History*, for February, 1839, and which is rarely found in the North American, or Old World Rodents. In enumerating the above groups, I omitted the *Caviidae*, because in the form of the lower jaw, they differ somewhat from the rest – they possess, in fact, a form of lower-jaw peculiar to themselves;

[29] I am acquainted with only one exception, and that is in the genus *Castor*. In the genus *Ondatra*, the descending ramus is but slightly twisted outwards, but in all the other *Arvicolidae*, whose crania I have examined, it is remarkably so, and in the genera *Spalax* and *Geomys*, where this character is carried to the extreme, the descending ramus projects from the alveolus of the long inferior incisors, in the form of a rounded and almost horizontal plate.

[30] In aged individuals of some of the species of *Arvicolidae*, the molar teeth possess short roots. In a skull of *Ondatra* now before me I find all the molars divided at the base into two portions, which in all probability would have formed solid roots had the animal lived longer.

but in the Chinchillas[31] the transitions between one form and the other are found.

The South American *Muridae*, which form the chief part of Mr Darwin's collection, were none of them procured further north than latitude 30°, with the exception of those from the Galapagos Archipelago. The species occur at the following localities.

SOUTH AMERICA

WEST COAST

GALAPAGOS ARCHIPELAGO
Mus Jacobiae
Galapagoensis

COQUIMBO
Mus longipilis
Renggeri
Darwinii

VALPARAISO
Mus Renggeri
decumanus

CONCEPCION
Mus longicaudatus

CHILOE AND CHONOS ARCHIPELAGO
Mus brachiotis

EAST COAST

MALDONADO
Mus *decumanus*
maurus
Musculus
tumidus
nasutus
obscurus
arenicola
bimaculatus
flavescens
Reithrodon typicus /

BUENOS AYRES
Mus *decumanus*

BAHIA BLANCA
Mus Braziliensis
elegans
gracilipes

RIO NEGRO
Mus griseo-flavus

PORT DESIRE
Mus canescens

ST JULIAN
Reithrodon cuniculoïdes
xanthopygus
Reithrodon cunicoloïdes

SANTA CRUZ
Mus canescens
micropus
xanthopygus
Reithrodon cuniculoïdes

FALKLAND ISLANDS
Mus *decumanus*
Musculus

STRAITS OF MAGELLAN
Mus xanthorhinus
Magellanicus
Reithrodon chinchilloïdes

[31] See *Proceedings of the Zoological Society* for 9 April, 1839, p. 61.

SECTION: HYSTRICINA

FAMILY: ------?

MYOPOTAMUS COYPUS

Myopotamus Coypus, *Auct.*

'This animal, in Chile, is known by the name of "Coypu"; at Buenos Ayres, where an extensive trade is carried on with their skins, they are improperly called "nutrias", or otters. In Paraguay, according to Azara, their Indian name is "guiya". On the east side of the continent they range from Lat. 24° (Azara) / to the Rio Chupat in 43° 20′; distance of 1,160 miles. This latter river is 170 miles south of the Rio Negro, and the intervening space consists of level, extremely arid, and almost desert plains, with no water, or at most one or two small wells. As the Coypu is supposed never to leave the banks of the rivers, and being, from its web-feet and general form of body, badly adapted for travelling on land, its occurrence in this river is a case, like so many others in the geographical distribution of animals, of very difficult explanation. The same remark is indeed applicable, but with less force, to its existence in the Rio Negro. On the west coast; it is found from the valleys of central Chile (Lat. 33°) to 48°S., or perhaps even somewhat farther, but not in Tierra del Fuego. So that, on the Atlantic side of the continent, the plains of Patagonia check its range southward, as, on the Pacific side, the deserts of Chile do to the north. Its range, including both sides, is from 24° to 48°, or 1,440 miles. In the Chonos Archipelago these animals, instead of inhabiting fresh water, live exclusively in the bays and channels which extend between the innumerable small islets of that group. They make their burrows within the forest, a little way above the rocky beaches. I believe it is far from being a common occurrence, that the same species of any animal should haunt indifferently fresh water, and that of the open sea. We shall see that the Capybara is sometimes found on the islands near the mouth of the Plata; but these cannot be considered as their habitual station in the same manner as the channels in the Chonos Archipelago are to the Coypu. The inhabitants of Chiloe, who sometimes visit this Archipelago for the purpose of fishing, state that these animals do not live solely on vegetable matter, as is the case with those inhabiting rivers, but that they sometimes eat shell-fish. The Coypu is said to be a bold animal, and to fight fiercely with the dogs employed in chasing it. Its flesh when cooked is white and good to eat. An old female procured (January) amongst these islands, weighed between ten and eleven pounds.' D.

FAMILY: OCTODONTIDAE

CTENOMYS BRAZILIENSIS

Ctenomys Braziliensis, *De Blainville, Bulletin de la Société Philomatique,* June, 1836, p. 62.

Maldonado, La Plata. (*June*)

'This animal is known by the name of Tucutuco. I have given an account of its habits in my journal, but I shall here repeat it for the sake of keeping / together my observations on the less known animals. The Tucutuco is exceedingly abundant in the neighbourhood of Maldonado, but it is difficult to be procured, and still more difficult to be seen, when at liberty. Azara,[32] who has given an account of its habits, with which every thing I saw perfectly agrees, states that he never was able to catch more than one, although they are so extremely common. The Tucutuco lives almost entirely under ground, and prefers a sandy soil with a gentle inclination; but it sometimes frequents damp places, even on the borders of lakes. The burrows are said not to be deep, but of great length. They are seldom open; the earth being thrown up at the mouth into hillocks not quite so large as those made by the mole. Considerable tracts of country are completely undermined by these animals. They appear, to a certain degree, to be gregarious; for the man who procured my specimens had caught six together, and he said this was a common occurrence. They are nocturnal in their habits, and their principal food is afforded by the roots of plants, which is the object of their extensive and superficial burrows. In the stomach of one which I opened I could only distinguish, amidst a yellowish green soft mass, a few vegetables fibres. Azara states that they lay up magazines of food within their burrows.

[32] Azara, *Voyages dans l'Amerique Meridionale,* vol. i, p. 324.

'The Tucutuco is universally known by a very peculiar noise, which it makes when beneath the ground. A person, the first time he hears it, is much surprised; for it is not easy to tell whence it comes, nor is it possible to guess what kind of creature utters it. The noise consists in a short, but not rough, nasal grunt, which is repeated about four times in quick succession; the first grunt is not so loud, but a little longer and more distinct than the three following: the musical time of the whole is constant, as often as it is uttered. The name Tucutuco is given in imitation of the sound. In all times of the day, where this animal is abundant, the noise may be heard, and sometimes directly beneath one's feet. When kept in a room, the Tucutucos move both slowly and clumsily, which appears owing to the outward action of their hind legs; and they are likewise quite incapable of jumping even the smallest vertical height. Mr Reid, who dissected a specimen which I brought home in spirits, informs me that the socket of the thigh-bone is not attached by a ligamentum teres; and this explains, in a satisfactory manner, the awkward movements of their hinder extremities. Their teeth are of a bright wax yellow, and are never covered by the lips: they are not adapted to gnaw holes or cut wood. When eating any thing, for instance biscuit, they rested on their hind legs and held the piece in their fore paws; they appeared also to wish to drag it into some corner. They were very stupid in making any attempt to escape; when angry or frightened, they uttered / the Tucutuco. Of those I kept alive, several, even the first day, were quite tame, not attempting to bite or to run away; others were a little wilder. The man who caught them asserted that very many are invariably found blind. A specimen which I preserved in spirits was in this state; Mr Reid considers it to be the effect of inflammation in the nictitating membrane. When the animal was alive, I placed my finger within half an inch of its head, but not the slightest notice was taken of it: it made its way, however, about the room nearly as well as the others. Considering the subterranean habits of the Tucutuco, the blindness, though so frequent, cannot be a very serious evil; yet it appears strange that any animal should possess an organ constantly subject to injury. The mole, whose habits in nearly every respect, excepting in the kind of food, are so similar, has an extremely small and protected eye, which, although possessing a limited vision, at once seems adapted to its manner of life.

'Several species probably will be found to exist south of the Plata. At Bahia Blanca (Lat. 39°) an animal burrows under ground in the same manner as the *C. Braziliensis*, and its noise is of the same general character, but instead of being double and repeated twice at short intervals, it is single and is uttered either at equal intervals, or in an accelerating order. I was assured by the inhabitants that these animals are of various colours, and, therefore, I presume that the two kinds of noises proceeded from two species. However this may be, they are extraordinarily numerous: many square leagues of country between the Sierras Ventana and Guetru-heigue are so completely undermined by their burrows, that horses in passing over the plain, sink, almost every step, fetlock deep. At the Rio Negro (Lat. 41°) some closely allied (or same?) species utters a noise, which is repeated only twice, instead of three or four times as with the La Plata kind. The sound is, moreover, louder and more sonorous; and so closely resembles that made in cutting down a small tree with an axe, that I have occasionally remained in doubt for some time to which cause to attribute it. Where the plains of Patagonia are very gravelly (as at Port Desire and St Julian) the Ctenomys, I believe, does not occur; but at Cape Negro, in the Strait of Magellan, where the soil is damper and more sandy, the whole plain is studded with the little hillocks, thrown up by this destructive animal. It occurs likewise south of the Strait, on the eastern side of Tierra del Fuego, where the land is level. Captain King brought home a specimen from the northern side of the Strait, which Mr Bennett[33] has called *C. Magellanicus*: it is of a different colour from the *C. Braziliensis*. I unfortunately did not make any note regarding the noise of this southern species: but the circumstance of its existence rather corroborates my belief in there being several other kinds in the neighbourhood of the Rio / Negro and Bahia Blanca. Otherwise we must believe that the same animal utters different kinds of noises, in different districts; a fact which I should feel much inclined to doubt.

'Azara[34] says that the Tucutuco may be "found every where; provided that the soil be pure sand, and the situation not subject to be overflowed. As these conditions are fulfilled only in certain spots, their warrens are far separated from each other, even sometimes more than twenty-five leagues, without it being possible to conceive how these animals have been able to pass from one place to another." The difficulty, I think, is much overstated; for, as I have said, the burrows of the Tucutuco are sometimes made in very damp places, near lakes; so that they certainly might pass over almost any kind of country. But

33 *Transactions of the Zoological Society*, vol. ii, p. 84.
34 Azara, *Voyage dans l'Amerique Meridionale*, vol. i, p. 324.

if the *C. Braziliensis* and *C. Magellanicus* be considered as one species, as some French authors are inclined to do, then the difficulty will be increased in a very remarkable manner, as we shall be obliged to transport the Tucutuco over wide plains of shingle, and across many great rivers, and an arm of the sea.' D

POEPHAGOMYS ATER

Poephagomys ater, *F. Cuvier, Annales des Sciences Naturelles*, 2d series, *Zoologie*, tom. 1, p. 321. June, 1834.

Chile. (*September*)

'This animal is generally scarce, but in certain districts, I believe, of an alpine character, it is abundant. It excavates very extensive superficial burrows, no doubt, for the purpose of feeding on the roots of plants, as in the case of the *Ctenomys Braziliensis*, the habits of which have just been described. Horses passing over districts frequented by these animals, sink fetlock deep through the turf. I procured my specimen from Valparaiso, where the country-people called it "Cururo".' D.

OCTODON CUMINGII

Octodon Cumingii, *Bennett, Proc. of Committee of Science and Correspondence of the Zool. Soc.* for 1832, p. 46.

Octodon Cumingii, *Transactions of the Zoological Society of London*, vol. ii, p. 81. Pl. 16.

Dendrobius Degus, *Meyen. Acta Academiae*, c. l. c. *Naturae Curiosorum*, xvi, p. 610. Pl. 44, 1833.

Valparaiso, Chile. (*October*) /

These little animals are exceedingly numerous in the central parts of Chile. They frequent by hundreds the hedge-rows and thickets, where they make burrows close together, leading one into another. They feed by day in a fearless manner; and are very destructive to fields of young corn; when disturbed, they all run together towards their burrows in the same manner that rabbits in England do when feeding outside a covert. When running they carry their tails high up, more like squirrels than rats; and they often remain seated on their haunches, like the former animals. According to Molina[35] they lay up a store of food for the winter, but do not become dormant. The Octodon is the 'degu' of that author: he says that the Indians in past times used to eat them with much relish. These animals appear to be very subject to be piebald and albinos; as if partly under the influence of domestication.

GENUS: ABROCOMA[36]

Dentes primores 2/2 acuti, eradicati, anticè laeves: molares utrinque 4/4 subaequales, illis maxillae superioris in areas duas transversales ob plicas vitreas acutè indentatus divisis; plicis utriusque lateris vix aequè profundis; illis mandibulae inferioris in tres partes divisis, plicis vitreis his internè, semel externè indentatis, areâ primâ sagittae cuspidem fingente, caeteris acutè triangularibus

Artus subaequales

Antipedes 4-dactyli, externo brevissimo, intermediis longissimis et ferè aequalibus

Pedes postici 5-dactyli; digito interno brevissimo. Ungues breves et debiles, illo digiti secundi lato et lamellari; omnibus setis rigidis obtectis

Caput mediocre, auribus magnis, membranaceis; oculis mediocribus

Cauda breviuscula

Vellus perlongum, et molle

The genus *Abrocoma* is evidently allied on the one hand to the genera *Octodon*, *Poephagomys*, and *Ctenomys*, and on the other to the family *Chincillidae*. The four genera just mentioned possess so many characters in common, that it would be well to unite them, and the name *Octodontidae* may be used to designate the group.

The *Octodontidae* appear to bear the same relations to *Echimys*, as the *Arvicolae* do to the *Muridae*. /

In the *Octodontidae* the skull is rather short, the inter-orbital space is broad; the ant-orbital passage is large; the zygomatic arch is thrown

[35] *Compendio de la Hist. Nat. del Reyno de Chile*, vol. i, p. 343.
[36] 'Αβρος, soft; Κομη, hair.

out horizontally from the plane of the palate; the malar bone is broad and somewhat compressed, and throws up a small post-orbital process; the glenoid cavity of the temporal bone is narrow; the palate is contracted, and deeply notched posteriorly, the portion which lies between the molar teeth descends below the level of the anterior portion; the incisive foramina are wide: the body of the anterior and posterior sphenoids is very narrow, and the foramina on either side of them are large: the occipital condyles are very narrow, widely separated, and the articular surface is nearly vertical.[37] The descending *ramus* of the lower jaw springs from the outer side of the alveolar portion, and terminates in a point, more or less acute.

The incisors of the upper and lower jaws are of the same width: the molars are $\frac{4-4}{4-4}$, rootless.

In external characters the species of the present group vary considerably. The toes are 5|5 or 4|5. The claws of the hind feet are covered by strong, curved bristly hairs.

The principal points of distinction in the external characters of the four genera under consideration, may be thus expressed.

TOES 5|5

A. Fore feet formed for burrowing – strong and armed with large claws; tail short
 a. Ears minute, incisors very broad — *Ctenomys*
 b. Ears small, incisors broad — *Poephagomys*

B. Fore feet weak; claws small; incisors narrow; ears large
 a. Tail with the apical portion furnished with long hair — *Octodon*

TOES 4|5

 b. Tail furnished throughout with short adpressed hairs — *Abrocoma*

It is not only in the comparatively small size and weakness of the fore feet that *Abrocoma* approaches more nearly to *Octodon*; but it agrees in having the soles, both of the fore and hind feet (which are devoid of hair), covered with minute round fleshy tubercles (see the under side of the tarsus figured in Plate 28).

In *Octodon*, however, the toes have on their under side transverse

[37] There is a wide difference between the present animals and the *Arvicolidae* in the form of the occipital condyles: the same difference is also observable between *Echimys* and *Mus*. The *Octodontidae* in fact have the same form of condyles as the Chinchillas and Cavies. In this and many other characters the last mentioned animals evince an affinity to the *Leporidae*.

incisions, as the *Muridae*, and many other Rodents; a character not found in *Abrocoma*. / Here the under-side of the toes, like the sole of the foot, is covered with minute tubercles.

Though in the form of the skull *Abrocoma Cuvieri*[38] agrees most nearly with that of *Octodon*; it differs in having the anterior portion narrower and rather larger, compared to the part devoted to the protection of the brain; the zygomatic arch is shorter, the incisive foramina are longer, the body of the anterior sphenoid is narrower, and the auditory bullae are larger. The principal differences observable in the form of the lower jaw of *Abrocoma*, when compared with that of *Octodon*, consists in the coronoid process being smaller, the condyloid narrower from front to back; the descending *ramus* more deeply emarginated posteriorly, and the angle longer and more attenuated.

In those characters in which the skull of *Abrocoma* departs from that of *Octodon*, it approaches nearer to *Chinchilla*. In the peculiar form and large size of the ears, in the extreme softness of the fur, in the greater development of the pads on the under side of the toes, and in the possession of only four toes to the fore feet, there are other points of resemblance between *Abrocoma* and *Chinchilla*. In the Chinchilla as well as in *Octodon* and *Abrocoma*, we find the toe corresponding to the second (counting from the inner side) furnished with a broad hollow nail;[39] there are also stiff bristly hairs covering this nail as in the *Octodontidae*.

The extreme softness of the fur of the animals about to be described, suggested for them the generic name of *Abrocoma*. The fur consists of hairs of two lengths, and the longer hairs are so extremely slender that they might almost be compared to the web of the spider. The specific names applied are those of the distinguished naturalists who first made us acquainted with the two genera, *Octodon* and *Poephagomys*.

[38] I have not had an opportunity of examining the skull of *Abrocoma Bennettii*.

[39] This nail no doubt is used to cleanse the fur, and the bristly hairs may also assist in the operation; the two small toes of the Kangaroo's hind foot are used for the same purpose.

1. ABROCOMA BENNETTII

PLATE XXVIII

Abrocoma Bennettii, *Waterh.*, *Proceedings of the Zoological Society of London*, for February, 1837, p. 31.

A. corpore suprà griseo, ad latera pallidiore et pallidè cervino lavato, subtùs albescenti-cervino; gulâ albescenti-griseâ; pedibus sordidè albis: auribus amplis, ad marginem posticum rectis, extùs ad bases vellere, sicùt in corpore, obsitis: caudâ corpore breviore, ad basin crassiusculâ, pilis brevibus incumbentibus vestitâ

Description. Form stout; ears large, with the posterior margin straight; fore / feet rather small, tarsus short; tail rather shorter than the body, thick at the base; fur long and extremely soft, and silk-like. General colour pale greyish brown, with a slight yellow wash; the upper part of the head and the back dusky brown; under parts of the body very pale yellowish brown, inclining to white; chin and throat whitish; feet dirty white; tail well clothed with hairs, which are closely adpressed, brown above, and of a very pale brown beneath at the base, darker towards the apex. The hairs of the moustaches are numerous, long, rather slender, and of a brownish colour. The ears are brown, furnished externally at the base with fur resembling that of the body; the remaining parts (both external and internal) are beset with long and extremely slender brown hairs, which project considerably beyond the margin of the ear. The ordinary fur on the back is about ten lines in length, but thickly interspersed with this fur, are longer hairs which are so delicate that they may almost be compared to the spider's thread. Both on the upper and under side of the body the fur is deep grey at the base. The incisors are yellow.

	Inches	*Lines*		*Inches*	*Lines*
Length from nose to root of tail	9	9	Length of tarsus (claws included)	1	4
of tail	5	0	of ear	0	10
from nose to ear	1	11	Width of ear	1	0½

Habitat, Chile. (*August*)

'This animal was caught among some thickets in a valley on the flanks of the Cordillera, near Aconcagua. On the elevated plain, near the town of Santa Rosa, in front of the same part of the Andes, I saw two others, which were crawling up an acacia tree, with so much facility, that this practice must be, I should think, habitual with them.' D.

2. ABROCOMA CUVIERI

PLATE XXIX

Abrocoma Cuvieri, *Waterh.*, *Proceedings of the Zoological Society of London* for February, 1837, p. 32.

Ab. suprà grisea, levitèr ochraceo lavata; abdomine gulâque albescenti-griseis; pedibus sordidè albis; auribus amplis, ad marginem posticum distinctè emarginatis; caudâ corpore multò breviore, et nigrescente

Description. Ears large; tail considerably shorter than the body; fur extremely / soft; general colour grey faintly washed with yellow; under parts of the body greyish white; feet dirty white; tail dusky, paler beneath at the base: the ears are large, distinctly emarginated behind, and appear to be almost naked, but, upon close examination, long and extremely fine hairs may be observed. All the fur is grey at the base; the hairs of the moustaches are numerous and very long, those nearest the mouth are white, the others are black at the base and greyish beyond. The incisors are of a palish yellow colour.

	Inches	Lines		Inches	Lines
Length from nose to root of tail	6	6	Length of tarsi (claws included)	1	1
of tail	2	10	of ear	0	7
from nose to ear	1	4	Width of ear	0	7½

Habitat, Chile. (*September*)

This species is about one-third the size of the last, it differs moreover in being grey instead of brown, and in having the posterior margin of the ear emarginated; the tail is also rather shorter in proportion.

The skull[40] is figured in Pl. XXXIII, fig. 1 *a* and 1 *b*; and fig. 23 *a*,

[40] The skull is, unfortunately, imperfect, the hinder portion is injured, and the arches which enclosed the ant-orbital openings are broken.

PLATE XXVIII

Abrocoma Bennettii

Pl. XXXIV. Its length is 1 inch, 4½ lines; width 9¼ lines; length of nasal bone 6 lines; distance between fore part of incisors and the front molar (upper jaw) 5 lines; longitudinal extent of the three molars of upper jaw 3 lines; length of auditory bullae 5¾ lines; length of *ramus* of lower jaw (see Pl. XXXIII, fig. 1 *c*), without incisors, 11½ lines. Fig. 23 *c*, Pl. XXXIV, represents the inner side of a *ramus* of the lower jaw: fig. 1 *d*, Pl. XXXIII, is the lower jaw seen from above: fig. 23 *b*, Pl. XXXIV, is the same seen from beneath. This view is given to show the position of the descending ramus of the lower jaw – that it springs from the outer side of the alveolar portion, as in a great portion of the South American Rodents, such as *Dasyprocta*, *Myopotamus*, *Echimys*, *Chinchilla*, and also in that genus found in the West Indian islands, *Capromys*. Fig. 1 *e*, Pl. XXXIII, represents the molar teeth of the upper jaw, and fig. 1 *f*, those of the lower.

'This species is abundant on the dry hills, partly covered with bushes, near Valparaiso.' D. /

FAMILY: CHINCHILLIDAE

LAGOSTOMUS TRICHODACTYLUS

Lagostomus trichodactylus, *Brooks*, *Transactions of the Linnean Society*, vol. xvi, p. 95, Pl. 9.

La Vizcache, *Azara*, *Essais sur l'Histoire Naturelle des Quadrupedes de la Province du Paraguay*, vol. ii, p. 41. Trad. Franc.

Vischacha, *Meyen*, *Acta Academie*, c. l. c. *Nature Curiosorum*, tom. xvi, pars 2, p. 584.

Habitat, La Plata.

'I will not repeat what I have said about the habits of this animal in my *Journal*, as it is merely a corroboration of Azara's account. According to that author, the Bizcacha is not found north of 30°; and its southern limit occurs in the neighbourhood of Rio Negro in 41°. Where the plains are gravelly, it is not abundant, but (differently from the *Cavia Patagonica*), it prefers an argillaceous and sandy formation, such as that near Buenos Ayres. The Bizcacha abounds over the whole Pampas, even to the neighbourhood of Mendoza, and there it is replaced in the Cordillera by an Alpine species. Of the latter animal, I saw one seated on a pinnacle at a great height, but I could not obtain a

PLATE XXIX

Abrocoma Cuvieri

specimen of it. Azara[41] has remarked that the Bizcacha, fortunately for the inhabitants of Banda Oriental, is not found to the eastward of the Rio Uruguay; and what makes the case more remarkable is, that although thus bounded by one river, it has crossed the broader barrier of the Parana, and is numerous in the province of Entre Rios. I was assured by a man, whose veracity I can perfectly trust, that these animals, quasi canes, post coitum adnexi sunt.' D.

FAMILY: CAVIIDAE

KERODON KINGII

Kerodon Kingii, *Bennett, Proceedings of the Zoological Society of London* for 1835, p. 190.

Habitat, Patagonia.

'The Kerodon is common at intervals along the coast of Patagonia, from the / Rio Negro (Lat. 41°) to the Strait of Magellan. It is very tame, and commonly feeds by day: it is said to bring forth two young ones at a birth. At the Rio Negro it frequents in great numbers the bottoms of old hedges: at Port Desire it lives beneath the ruins of the old Spanish buildings. One old male killed there weighed 3,530 grains. At the Strait of Magellan, I have seen amongst the Patagonian Indians, cloaks for small children made with the skins of this little animal; and the Jesuit Falkner says, that the people of one of the southern tribes, take their name from the number of these animals which inhabit their country. The Spaniards and half-civilized Indians, call the Kerodon, "conejos", or rabbit; and thus the mistake has arisen, that rabbits are found in the neighbourhood of the Strait of Magellan.' D.

1. CAVIA COBAIA

Cavia Cobaia, *Auct.*

Habitat, Maldonado, La Plata. (*June*)

'This animal, known by the name of Aperea, is exceedingly common in

[41] Azara, *Voyages dans l'Amerique Meridionale*, vol. i, p. 316.

the neighbourhood of the several towns which stand on the banks of the Rio Plata. It frequents different kinds of stations – such as hedge-rows made of the Agave and Opuntia, or sand-hillocks, or again, marshy places covered with aquatic plants; the latter appearing to be its favourite haunt. Where the soil is dry, it makes a burrow; but where otherwise, it lives concealed amidst the herbage. These animals generally come out to feed in the evening, and are then tame; but if the day be gloomy, they make their appearance in the morning. They are said to be very injurious to young trees. An old male killed at Maldonado, weighed 1 lb 3 oz. In all the specimens I saw there (during June, or winter), I observed, that the hair was attached to the skin less firmly than in any other animal I remember to have seen.' D.

2. CAVIA PATACHONICA

Cavia Patachonica, *Shaw, General Zoology*, vol. ii, part 1, p. 226.
Dasyprocta Patachonica, *Desmarest, Mamm.*, p. 358, Sp. 574.
Dolichotis Patachonica, *Desmarest, Mamm.*, in Note, p. 359–60.
Chloromys Patachonicus, *Lesson, Manuel de Mammalogie*, p. 301.
Lièvre Pampa, *Azara, Essais sur l'Histoire Naturelle des Quad. de la Province du Paraguay.* French Translation, vol. ii, p. 51.

In the form of the cranium, and in the structure of the teeth, this animal possesses all the characters of the Cavies (*Caviidae*).[42]

Habitat, Patagonia. /

'This animal is found only where the country has rather a desert character. It is a common feature in the landscape of Patagonia, to see in the distance two or three of these Cavies hopping one after another in a straight line over the gravelly plains, thinly clothed by a few thorny bushes and a withered herbage. Near the coast of the Atlantic, the northern limit of this species is formed by the Sierra Tapalguen, in latitude 37° 30', where the plains rather suddenly become greener and more humid. The limit certainly depends on this change, since near Mendoza (33° 30'), four degrees further northward, where the country is very sterile, this animal again occurs. Azara erroneously

[42] See *Proceedings of the Zoological Society* for April, 1839, p. 61.

supposed that its northern range was only 35°.[43] It is not clear on what circumstances its limit southward between Ports Desire and St Julian (about 48° 30′) depends; for there is in that part no change in the features of the country. It is, moreover, a singular circumstance, that although the Cavy was not seen at Port St Julian during our voyage, yet Captain Wood, in 1670, speaks of them as being numerous there. What cause can have altered, in a wide, uninhabited, and rarely visited country, the range of an animal like this?

'Azara states,[44] that the Cavy never excavates its own burrow, but uses that of the Bizcacha. Wherever this animal is present, without doubt this is true; but on the sandy plains of Bahia Blanca, where the Bizcacha is not found, the Spaniards maintain that the Cavy is its own workman. The same thing occurs with the little owls of the Pampas (*Noctua cunicularia*), which have been described by travellers as standing like sentinels at the mouths of almost every burrow; for in Banda Oriental, owing to the absence of the Bizcacha, these birds are obliged to hollow out their own habitations. Azara says, also, that this Cavy, except when pressed by danger, does not enter its burrow; on this point I must again differ from that high authority. At Bahia Blanca I have repeatedly seen two or three of these animals sitting on their haunches by the mouths of their holes, which they quietly entered as I passed by at a distance. Daily, in the neighbourhood of these spots, the Cavies were abundant: but differently from most burrowing animals, they wander, commonly two or three together, to miles or leagues from their home; nor do I know whether they return at night. The Cavy feeds and roams about by day; is shy and watchful; seldom squats after the manner of a hare; cannot run very fast, and, therefore, is frequently caught by a couple of dogs, even of mixed breed. Its manner of running more resembles that of a rabbit than of a hare. The Cavy generally produces two young ones at a birth, which are brought forth within the burrow. The flesh, when cooked, is / very white; it is, however, rather tasteless and dry. Full grown animals weigh between twenty and twenty-six pounds.' D.

43 Azara, *Voyage dans l'Amerique Méridionale*, vol. i, p. 318.
44 Azara, *Quadrupeds of Paraguay*.

HYDROCHOERUS CAPYBARA

Hydrochoerus Capybara, *Auct.*

'These animals are common wherever there are large rivers or lakes, over that part of the South American Continent which lies between the Orinoco and the Plata, a distance of nearly 1,400 miles. They are not generally supposed to extend south of the Plata; but as there is a Laguna Carpincho (the latter being the provincial name of the Capybara) high up the Salado, I presume they have sometimes been seen there. Azara does not believe they ever frequent salt water; but I shot one in the Bay of Monte Video; and several were seen by the officers of the *Beagle* on the Island of Guritti, off Maldonado, where the water is very nearly as salt as in the sea. The one I shot, at Monte Video, was an old female; it measured from tip of snout to end of stump-like tail, 3 feet 8½ inches, and in girth 3 feet 2 inches. She weighed 98 pounds. I opened the stomachs of a couple, which I killed near a lake at Maldonado, and found them distended with a thin yellowish-green fluid, in which not more than a trace of a vegetable fibre could be distinguished: it is in accordance with this fact, that a part of the aesophagus is so narrow, as I am informed by Mr Owen, that scarcely anything larger than a crow-quill can be passed down it. The shape of the dung of these animals is a short straight cylinder, rounded at the extremities; when dried and burnt, it affords a pleasant smell like that from cedar wood. These animals do not burrow holes, but live among the thickets, or beds of rushes near rivers and lakes. At Maldonado they often may be seen during the day, seated on the grassy plain in small groups of three and four, at the distance of a few yards from the border of the lake, which they frequent. I must refer the reader for a few more details respecting their habits, to my *Journal of Researches*. D.' /

SECTION: LEPORINA

FAMILY: LEPORIDAE

LEPUS MAGELLANICUS

Lepus Magellanicus, *Lesson et Garnot, Zoologie du Voyage autour du Monde de la Corvette,* La Coquille.

'A black variety of the domesticated species, which was turned out on these islands by the earlier colonists, has been considered, but with some hesitation, by M. Lesson, as a distinct species. He has called it *Lepus Magellanicus*, and has given the following specific character – '*Pilis omnino atro-violaceis, albis passim sparsis: auriculis fuscis, capite brevioribus; maculâ albâ naso, interstitio narium, menti, gulae, frontique.*'[45] In the specimens preserved on board the *Beagle*, the form and position of the white marks neither agree with M. Lesson's description, nor with each other. In one there is a broad white patch on one side of the head, and another on one of the hinder thighs. The Spaniards employed in hunting wild cattle (who are all excellent practical observers), assured me, that the black rabbits were only varieties of the common grey kind, and they gave the following reasons for thinking so; namely, that the two sorts did not live apart; that the black one had not a different range from the other; that the two bred freely together, and that they produced pie-bald offspring. As the rabbits extend their range very slowly (not having yet crossed the central range), the Spaniards have sometimes carried a few and turned them out in different parts of the island, and thus they have ascertained that the black and grey kinds breed together freely. Bougainville, moreover, who visited the part of the island, where the black variety is now most common, distinctly states, in his voyage round the world, that no animal, excepting the great wolf-like fox inhabited these islands. M. Lesson supposes that the *Lepus Magellanicus* is found near the Strait of Magellan; but I inquired of the Indians, who live there, and they knew of no other "conejos" or rabbits, except the *Kerodon Kingii*, which no doubt is the animal alluded to by the early voyagers.' D.

[45] *Voyage de* La Coquille. *Partie Zoologique*, vol. i, p. 168

ORDER: EDENTATA

1. DASYPUS HYBRIDUS

Dasypus hybridus, *Auct.*

'This species seems to prefer rocky and slightly undulating ground, and / hence is common in Banda Oriental and Entre Rios. Azara says it is found from 26° 30′, to at least 41° south; but, I was assured, perhaps incorrectly, that the Sierra Tapalguen (37° 30′), where the nature of the country becomes slightly different, is its southern limit. The *D. villosus*, *minutus*, and *mataco*, are found at Bahia Blanca, in latitude 39°. I was also assured that these three species, together with the *D. hybridus*, frequent the plains near Mendoza, in latitude 33° to 34°.' D.

2. DASYPUS MINUTUS

Dasypus minutus, *Auct.*

'The northern limit of this species on the Atlantic side of the continent, is (as I was told by the inhabitants) near the southern one of the *D. hybridus*, namely, 37° 30′. It is extremely abundant on the arid plains near the Sierra Ventana, and likewise in the neighbourhood of the Rio Negro. This species has a range considerably further southward than any other: I obtained specimens at Port Desire, where, however, it is far from common, and at Santa Cruz (in latitude 50°) I saw its tesselated covering lying on the ground. At Bahia Blanca, I found in the stomach of this armadillo, coleoptera, larvae, roots of plants, and even a small snake of the genus Amphisbaena. All the species, excepting one, wander about by day. At Bahia Blanca, during a morning's ride, three or four of the *D. minutus* generally were met with; but, in order to secure them, it was necessary to jump off one's horse as quickly as possible, otherwise, they would have disappeared by burrowing in the sandy soil. This species often endeavours to escape detection by squatting close to the ground, and remaining motionless.' D.

ORDER: MARSUPIALIA

1. DIDELPHIS AZARAE

Didelphis Azarae, *Auct.*

'This species is said to inhabit burrows: it is nocturnal, and is very destructive to poultry. The body after death possesses a very offensive odour. My specimen was procured at Maldonado.' D. /

2. DIDELPHIS CRASSICAUDATA

PLATE XXX

Didelphis crassicaudata, *Desmarest, Nouv. Dict. d'Hist. Nat.* 2d Ed., ix, p. 425.
Didelphis crassicaudata, *Mammalogie*, p. 257, Species 393.
Microuré troisième, ou Macrouré à gross queue, *Azara, Essais sur l'Histoire Nat. des Quad. de la Province de Paraguay*, vol. i, p. 284.

D. capite brevi; auribus parvis; colore corporis fuscescenti-flavo subtùs pallidiore; infra oculos flavescente; caudâ crassâ, caput corpusque, quoad longitudinem, ferè aequante; ad basin corporis colore tinctâ, dein nigra, ad apicem albâ

Description. Head short; ears small, the posterior edge emarginated near the base, distinctly furnished with hairs; tail slightly exceeding the body in length, very thick at the base; tarsi small; fur moderately long, slightly harsh, and somewhat adpressed (much less woolly than in most Opossums): general tint brownish yellow, under parts paler; anterior angle of the eye and muzzle brown, the tip of the chin, and also the tip of the muzzle on either side whitish; on the cheeks, a little below the eyes, is a patch of yellow which extends round the angle of the mouth: about one-third of the tail is covered with fur of the same colour and character as that on the body; beyond this the tail is black, excepting a small portion, about one inch in length, at the apex, which is white; and the hairs are short, closely adpressed, and scarcely hide the scales which are beneath: the fore portion of each foot is brown: the hairs covering the ears on the outer side are brownish, and those on the

PLATE XXX

Didelphis crassicaudata

inner side of the ear are yellow, but towards the outer margin they are brown. The hairs of the back have the basal half grey, and the apical half ochreous, terminating in yellowish brown; on the belly and underside of the neck, the hairs are ochreous, faintly tinted with grey at the base.

	Inches	*Lines*		*Inches*	*Lines*
Length from nose to root of tail	11	3	Length from nose to ear	2	1½
of tail	10	3	of tarsus	1	5½
			of ear	0	6

Habitat, Maldonado, La Plata. (*June*)

The species described by Azara, under the name *Macrouré à grosse queue*, agrees so perfectly with the present animal, that I have no hesitation in referring / it to the *Didelphis crassicaudata* of Desmarest, which is founded upon Azara's description.

The head of the *Didelphis crassicaudata* is shorter and less pointed than in most other Opossums; the ears are unusually small, and the tail is very thick. In the character of the fur also, this species differs from most others, the hairs being rather short and somewhat adpressed; and the soft under-fur being very scanty. Upon separating the fur on the back and sides of the body, numerous young hairs were visible in the specimen from which the above description is taken, and these were of a bright rusty red tint; the colouring of the animal therefore would, in all probability, have been very different after a short time, had it not been killed. Those observed by Azara varied considerably in their colouring. The skull is figured in Pl. XXXIV, figs 25. Fig. *d* represents a *ramus* of the lower jaw.

	Inches	*Lines*
Length of the skull	2	4
Width	1	3
Length of nasal bones	0	9½
Length of palate	1	2¾
Width of palate between the posterior molars	0	5
Distance between forepart of front incisors and forepart of canine	2	0¾
Distance between forepart of canine and hinder part of last molar	1	0
Length of *ramus* of lower jaw (incisors not included)	1	10½

'This specimen was caught at Maldonado: it weighed 14½ oz.' D.

3. DIDELPHIS ELEGANS

PLATE XXXI

D. vellere longo et molli, corpore suprà cinereo-fuscescente lavato; pedibus corporeque subtùs albis, oculis nigro circumdatis, interspatio cinerescente; auribus magnis fuscescentibus; caudâ, capite et corpore, paulo breviore

Description. Muzzle slender and pointed; ears large; tail rather shorter than the head and body taken together; fur long and very soft; general tint of the upper parts of the head and body ashy-grey washed with brown; on the sides of the body, especially near the shoulders, a faint yellowish tint is observable; the lower part of the cheeks, the throat, under parts of the body and the feet, are white, with an indistinct yellowish tint; the eyes are encircled with brownish-black, which tint is extended forwards on to the sides of the muzzle; the upper surface of the muzzle and the inter-orbital space is / pale. The tail is furnished throughout with minute decumbent hairs, excepting a small naked space at the tip beneath, of about one line in length; on the upper surface they are brown, and on the under, they are whitish. The fur of the upper and under parts of the body is deep grey at the base; on the lower part of the cheeks, chin, and on the mesial line of the throat and chest, the hairs are uniform – not grey at the base. The ears are brown, and to the naked eye, appear naked.

	Inches	*Lines*
Length from nose to root of tail	4	6
of tail	4	4
of tarsus (claws included)	0	7½
Length from nose to ear	1	1½
of ear	0	7¼
width of ear	0	7½

Habitat, Valparaiso, Chile. (*October*)

This little Opossum, which is the only species I am acquainted with from the west side of the Cordillera, was exhibited at one of the scientific meetings of the Zoological Society, and its characters were pointed out by Mr James Reid, who proposed for it the specific name of *hortensis*,[46] a name which was given from the circumstance that in

[46] See *Proceedings of the Zoological Society of London* for January, 1837, p. 4; its characters were not published.

Mr Darwin's notes it is stated that a small Opossum was found in a garden at Maldonado. These notes however refer to the *Didelphis brachyura*. The skull of this animal is figured in Pl. XXXV. Fig. 5 *a*, represents the upper side; 5 *b*, the under side; and 5 *c*, is the side view. Fig. 5 *d*, is the lower jaw, and 5 *e*, is the same magnified. The length of the skull is 14½ lines; width, 8 lines; length of palate, 7¼ lines; inter-orbital space, 2½ lines; length of *ramus* of lower jaw, 10½ lines. In the palate are two long openings which commence opposite the posterior false molar, and terminate opposite the hinder portion of the penultimate true molar: the incisive foramina are nearly one line in length. On the posterior portion of the palate there are four other foramina, one on each side near the posterior molar, and one on either side the mesial line, behind the large palatine openings above mentioned.

'These little animals frequent the thickets growing on the rocky hills, near Valparaiso. They are exceedingly numerous, and are easily caught in traps baited either with cheese or meat. The tail appeared to be scarcely at all used as a prehensile organ; they are able to run up trees, with some degree of facility. I could distinguish in their stomachs the larvae of beetles.' D. /

4. DIDELPHIS BRACHYURA

PLATE XXXII

Didelphis brachyura, *Auct.*

D. vellere brevi, corporis suprà cinereo, flavo lavato; lateribus capitis, corporisque, et partibus inferioribus rufescenti-flavis, gulâ et abdomine pallidioribus; caudâ brevi

Description. Head large; canine teeth very large; ears rather small; tail short, rather more than half the length of the body; fur short and crisp; the back and upper surface of the head ashy grey, grizzled with yellowish white; the sides of the head and body, and under parts rusty yellow, rather paler on the belly than on other parts, and of a deeper hue on the rump and cheeks; the eye is encircled with rusty yellow; feet yellowish; tail clothed with short stiff hairs, and exhibiting scales,

PLATE XXXI

Didelphis elegans

brownish above, and dirty yellowish white beneath – a small naked space beneath, at the tip, of about two lines in length. Fur of the back greyish at the base, that on the belly uniform; ears clothed with minute yellowish white hairs.

	Inches	*Lines*		*Inches*	*Lines*
Length from nose to the root of tail	6	0	Length of tarsus (claws included)	0	8¾
from nose to ears	1	6	of ear	0	3¾
of tail	2	8			

Habitat, Maldonado, La Plata. (*June*)

Never having seen a good figure of this animal, I have thought it desirable to introduce it in the plates of this work.

The *Didelphis brachyura* is closely allied to the *D. tricolor* of authors, but in that species the upper parts of the body are nearly black; the sides of the head and body are of a deep rusty red tint, and the under parts are almost white.

'Was caught by some boys digging in a garden. Its intestines were full of the remains of insects, chiefly ants and others of the Hemipterous order.' D./

PLATE XXXII

Didelphis brachyura

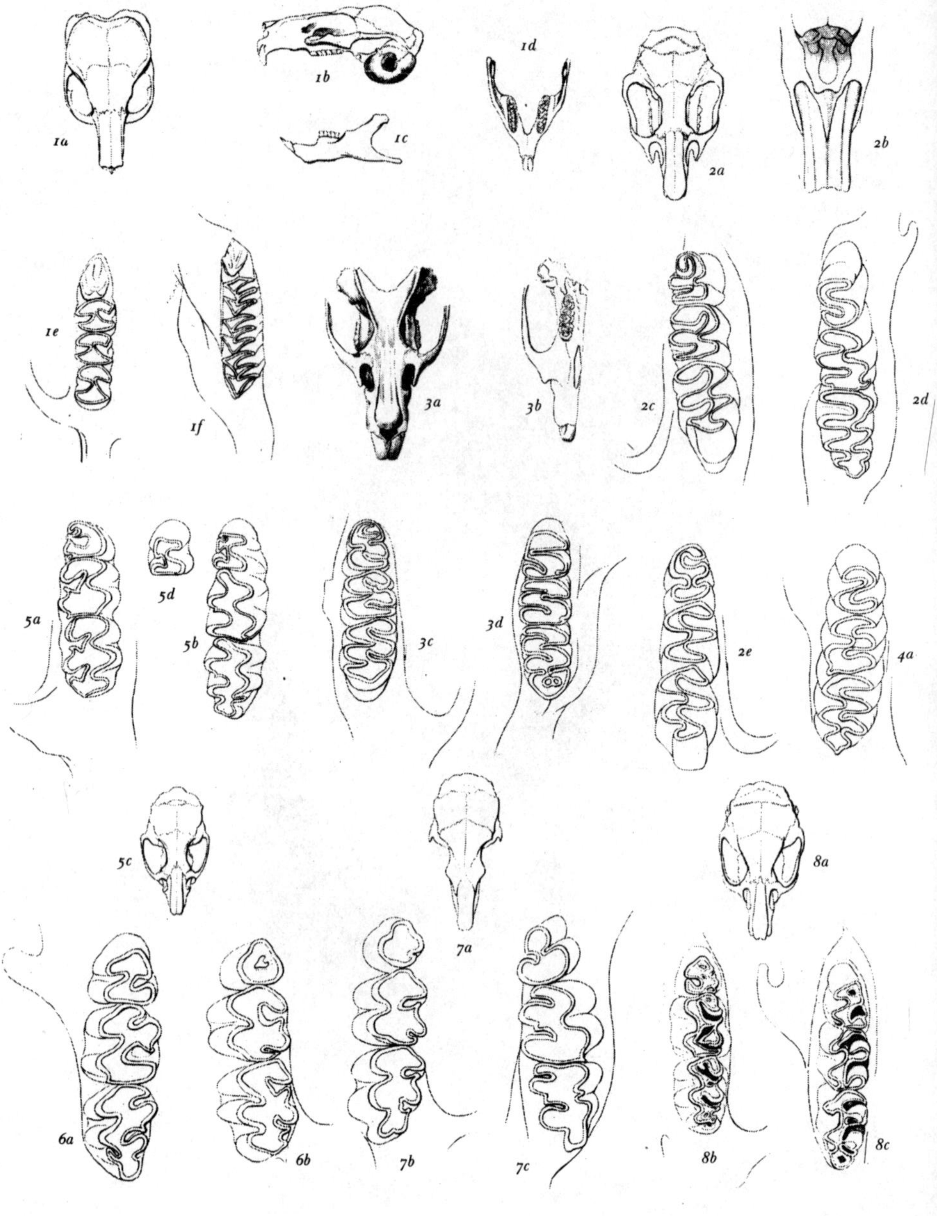

PLATE XXXIII

Skulls and molar teeth of various species of rodents

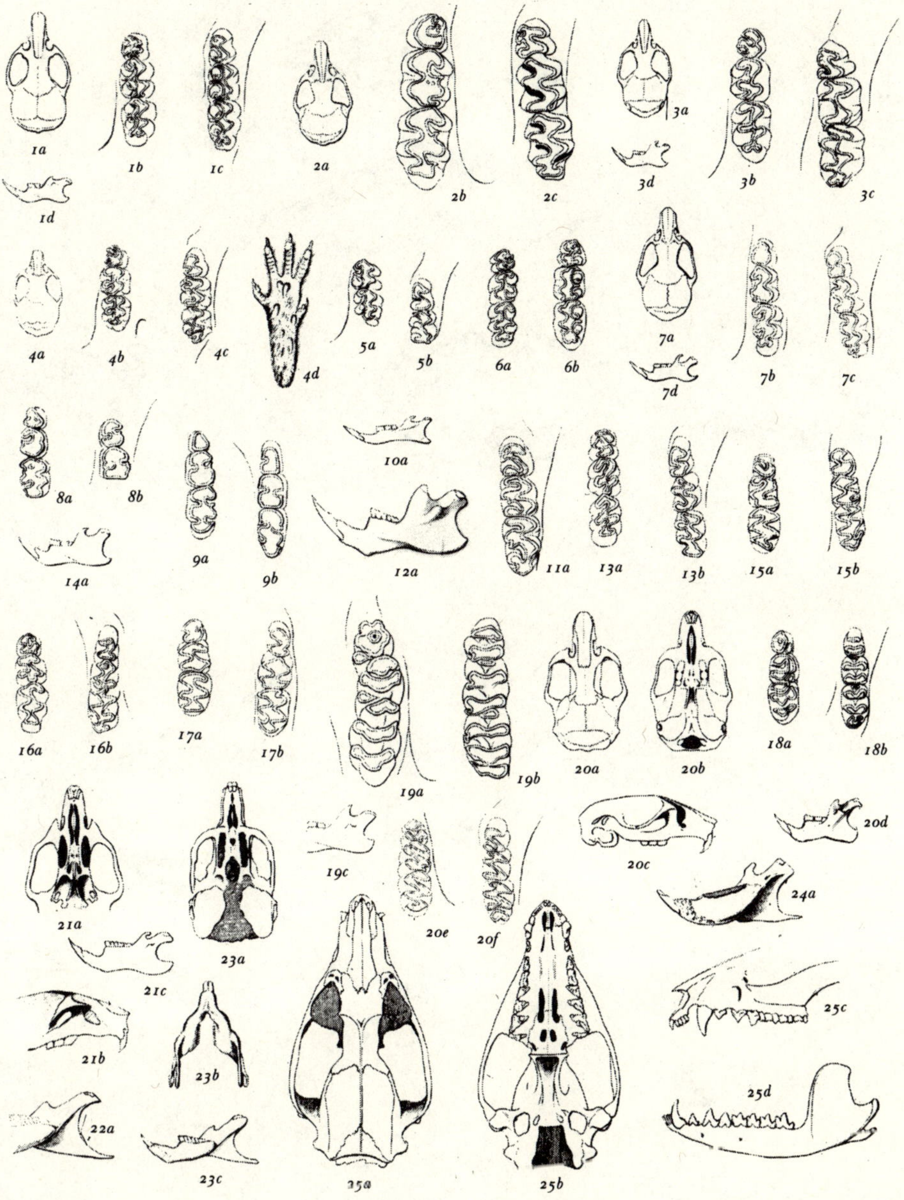

PLATE XXXIV

Skulls and molar teeth of various species of rodents, etc.

PLATE XXXV

Skulls of various animals